The Command Companion of Seamanship Techniques

Additional works by the same author

Seamanship Techniques (Combined Volume Parts 1 & 2), ISBN 07506 4135 5
Cargo Work (Kemp & Young) (6th Edition revised), ISBN 07506 3988 1
Marine Survival and Rescue Systems, ISBN 18560 9127 9
Helicopter Operations at Sea – A Guide to Industry, ISBN 18560 9168 6
Navigation for Masters, ISBN 18560 9147 3

The Command Companion of Seamanship Techniques

Volume 3

D. J. HOUSE

LONDON AND NEW YORK

First published by Butterworth-Heinemann

First published 2000

This edition published 2011 by Routledge
2 Park Square, Milton Park, Abingdon, Oxon OX14 4RN
711 Third Avenue, New York, NY 10017, USA

Routledge is an imprint of the Taylor & Francis Group, an informa business

British Library Cataloguing in Publication Data
A catalogue record for this book is available from the British Library

Library of Congress Cataloguing in Publication Data
A catalogue record for this book is available from the Library of Congress

ISBN-13: 978-0-750-64443-3

Contents

Introduction

The purpose of this companion volume is twofold, first to refresh age-old seamanship practice and secondly to update the serving Master with some of the many changes that are affecting the marine industry today. The ideas expressed are not meant to displace accepted methods, but more to show alternative avenues of approach.

With the massive increase in communications, information technology and major advances in electronic navigation systems it would be remiss if education providers and learning resources did not attempt to keep our serving Masters and Watch Officers abreast of our changing times. The electronic chart is rapidly eliminating the laborious task of individual chart correcting, while GPS and weather satellite systems have changed our very working concepts. Not only are these advances providing greater continual accuracy but, comparatively speaking, have made the Mariner's task that much easier.

New ideas in marine engineering with increased manoeuvring aids have changed ships' speeds, closing distances and anti-collision thinking. This is especially so with more and more tonnage being fitted with bridge control systems. The need for Masters to keep pace with change and also with the up-and-coming young officer with enhanced computer skills has become an essential element of the profession. However, the prime need for practical seamanship has in no way diminished, but a more modernized approach may be more appropriate for the new millennium.

D. J. House

Acknowledgements

Bell Helicopters USA
Blackpool & Fylde College, Fleetwood Nautical Campus
Butterworth Systems (UK) Ltd
Cable & Wireless Global Marine
Dunlop Beaufort Canada
Director General for Port State Control, Paris
E.H. Industries Ltd
Elliot White Gill
Furuno Electric (UK) Ltd
Her Majesty's Stationery Office.
HM Customs & Excise
Holland Roer-Propeller
International Salvage Union
J.M. Voith GmbH, Ship Techn. Div.
Kelvin Hughes Ltd
Kort Propulsion
Liferaft Systems Australia
Lloyd's (London)
Marine Accident Investigation Branch
Maritime and Coastguard Agency
Ministry of Transport/Director-General for Freight Transport, The Hague
Piracy Reporting Centre, ICC-International Maritime Bureau
Raytheon Marine GmbH
Salvage Association
Shell International Trading & Shipping Company

Smit Americas Inc.
Smit Nederland BV
Stena Sealink HSS
STN Atlas Electronik GmbH
United Technologies, Sikorsky Aircraft
US Coastguard
Viking Life Saving Equipment Ltd.
Westland Group plc
Witherby & Co. Publishers

Additional photographic work by

Mr George Edwards, Marine Chief Engineer (Rtd)
Captain Keith Millar, Lecturer Nautical Studies
Captain Sydney Garratt, Lecturer Nautical Studies
Mr Roger Smith
Mr H. Sason, Chief Officer MN
Mr Kevin Lloyd, Chief Officer MN

Additional thanks for the co-operation of

Mr Alan Witherby (Director) Witherby & Company Marine Publishers
The College Research Group of the Blackpool & Fylde College
The Fleetwood Offshore Survival Centre

Cover photograph supplied by Christopher David House.

List of abbreviations

AEW	Airborne Early Warning
AIS	Automatic Identity System
AKD	Auto Kick Down
ARCS	Admiralty Raster Chart Service
ARPA	Automatic Radar Plotting Aid
ASW	Anti Submarine Warfare
Aux	Auxiliary
BA	Breathing Apparatus
B/L	Bill of Lading
BP	British Petroleum
CBT	Clean Ballast Tank
CCTV	Close Circuit Television
CD	Compact Disc
CGSA	Carriage of Goods by Sea Act
c.i.f.	cost insurance and freight
CNIS	Channel Navigation Information Service
CO_2	Carbon Dioxide
COW	Crude Oil Washing
C/P	Charter Party
CPP	Controllable Pitch Propeller
CPSC	Certificate of Proficiency in Survival Craft (see PSCARB)
CSP	Commencement Search Pattern
CSWP	Code of Safe Working Practice
DETR	Department of Environment, Transport Regions
DF	Direction Finder
DGPS	Differential Global Positioning System
DOC	Document of Compliance
DP	Dynamic Positioning
DPOs	Dynamic Position Officers
DSC	Digital Selective Calling
DSV	Diving Support Vessel
DV	Desired Value
E	Error signal
ECDIS	Electronic Chart and Data Information Service

EDH	Efficient Deck Hand
ENC	Electronic Navigation Chart
EPIRB	Electronic Position Indicating Radio Beacon
FM	Frequency Modulation
f.o.b.	free on board
FOSC	Fleetwood Offshore Survival Centre
FOSC (US)	Federal On-Scene Commander
FRC (FRB)	Fast Rescue Craft (Boat)
FSU	Floating Storage Unit
FVQ	Flexible Vocational Qualification
FWE	Finished With Engines
GM	Metacentric Height
GMDSS	Global Maritime Distress and Safety System
GMP	Garbage Management Plan
GPS	Global Positioning System
GT	Gross Tonnage
HF	High frequency
HM	Her Majesty's
HMSO	Her Majesty's Stationery Office
HND	Higher National Diploma
HSC	High Speed Craft
IAMSAR	International Aeronautical and Maritime Search and Rescue Manual
IAPPC	International Air Pollution Prevention Certificate
IFR	Instrument Flying Rating
IG	Inert Gas
ILO	International Labour Organization
IMDG	International Marine Dangerous Goods (code)
IMO	International Marine Organization
INLS	International Pollution Prevention Certificate for Noxious Liquid Substances (in bulk)
IOPP	International Oil Pollution Prevention (certificate)
ISM	International Safety Management (code)

ISPPC	International Sewage Pollution Prevention Certificate		OPRC	Oil Pollution Preparedness Response and Co-ordination
ISU	International Salvage Union		O/S	Offshore
			OSC	On-Scene Commander
km	Kilometre		OSV	Offshore Supply Vessel
kW	Kilowatt			
			P & A	Procedures and Arrangement (manual)
LOF	Lloyd's Open Form (salvage)			
LPG	Liquid Petroleum Gas		PI	Parallel Indexing
LR	Lloyd's Register		P & I	Protection and Indemnity Club
			PL	Position Line
MAIB	Marine Accident Investigation Branch		PPM	Parts per Million
			PRS	Position Reference System
MAREP	Marine Reporting System (English Channel)		PSC	Port State Control
			PSCARB	Proficiency in Survival Craft and Rescue Boats
MARPOL	International Convention for the Prevention of Pollution			
MC	Medical Care		RAF	Royal Air Force
MCA (ex MSA)	Maritime and Coastguard Agency		RBD	Return of Births and Deaths
			RCC	Rescue Co-ordination Centre
MEC	Marine Evacuation Chute		RCDS	Raster Chart Display System
MERSAR	Merchant Ship Search and Rescue Manual (now IAMSAR)		RoT	Rate of Turn
			ROV	Remote Operated Vehicle
MES	Marine Evacuation System			
MF	Medium Frequency		SALS	Single Anchor Leg System
MGNs	Marine Guidance Notices		SAR	Search and Rescue
MINs	Marine Information Notices		SBM	Single Buoy Mooring
MMSI	Maritime Mobile Service Identity		SBP	Shore Based Pilot
MNTB	Merchant Navy Training Board		SBV	Standby Vessel
MoB	Man Over Board		SI	Statutory Instrument
MOU	Memorandum of Understanding		SMC	Safety Management Certificate
MPCU	Marine Pollution Control Unit		SMS	Safety Management System
MRCC	Marine Rescue Co-ordination Centre		SOLAS	Safety of Life at Sea Convention
MSI	Mobile Service Identity		SOPEP	Ships Oil Pollution Emergency Plan
MSNs	Marine Shipping Notice (previously 'M' Notice)		SQA	Scottish Qualification Authority
			SMC	Safety Management Certificate
MT	Motor Transport		SMS	Safety Management System
MV	Motor Vessel		STCW (convention)	Standards of Training, Certification and Watchkeeping
MV	Measured Value			
MVO	Merchant Vessel Operations		SWL	Safe Working Load
			SWOPS	Single Well Oil Production System
NINAS	Nucleus Integrated Navigation System			
			TAF-F	Training to Advanced Fire-fighting
NP	National Publication			
NSFCC	National Strike Force Co-ordination Centre		UHF	Ultra High Frequency
			UKC	Under Keel Clearance
NUC	Not Under Command		UKOPP	United Kingdom Oil Pollution Prevention (Certificate)
NVQ	National Vocational Qualification			
			UKSTC	United Kingdom Standard Conditions
OLB	Official Log Book			
OOW	Officer of the Watch		ULCC	Ultra Large Crude Carrier
OPA	Oil Pollution Act		UN	United Nations
OPIC	Oil Pollution Insurance Certificate		UMS	Unmanned Machinery Space

US	United States	VQ	Vocational Qualifications
USN	United States Navy	VRM	Variable Range Marker
		VTS	Vessel Traffic Scheme
VAT	Value Added Tax		
VHF	Very High Frequency		
VLCC	Very Large Crude Carrier	WRE	Wreck

1 A ship Master's business

A ship's Master is more and more being described as the ship's manager, where he or she in this modern day can expect to carry out the day-to-day, and the voyage-to-voyage, business of operating the ship not only safely but effectively in economic terms. Clearly the start of business will commence when a new Master takes over either a new ship, or relieves the Master from a previous voyage.

The usual pattern of handover documents and certificates is the norm and will depend on the type of ship for which the command is being assumed. The following list expands briefly on each of the documents.

Master's handover documents

1. The Certificate of Registry
   The Certificate of Registry contains the ship's name and the Port of Registry together with the ship's details, type, tonnage, engines, dimensions, builders etc. It will also contain the first Master's name and certificate number, in addition to the owner's name and address with details of any shared ownership. (NB. There are 64 shares in a ship and a company normally registers as the owner of a share or shares. Anyone who owns more than 33 shares becomes the majority shareholder.)

2. The Official Log Book (OLB)
   This is issued by a Port Superintendent or proper officer to the ship's Master, with the Crew Agreement and the ship's Articles, opened at the beginning of the voyage. Once a new Master relieves the existing Master or officer in charge and 'handover' is complete a statement to this effect is entered into the OLB.

3. Loadline Certificate (period of validity 5 years, may be extended 5 months)
   It is the Master's responsibility not to take the ship to sea in a condition in which she is overloaded. No ship is allowed to go to sea unless she has been surveyed and her deck and loadlines marked with the conditions of assignment. The Master will also be provided with stability, loading and ballast information in order to comply with the requirements of the Loadline Rules.

4. Safety Equipment Certificate (and record of inspection) (validity 2 years)
   A relieving Master should sight the validity of the Safety Equipment Certificate and ensure that it remains valid for the period of the voyage. In the event of expiry such arrangements must be made to conduct a safety equipment survey with the Marine Authority.

5. Cargo Ship Safety Construction Certificate (validity 5 years)
   This certificate is issued by or on behalf of the government agency of the country in which the ship is registered. Prior to the issue of this certificate the ship must be fitted with a type tested compass (no extension period) – a Passenger Ship Safety Certificate (equivalent) has a validity period of 1 year.

6. GMDSS Safety Radio Certificate (validity 1 year)
   By 1 February 1999 all ships at sea must be fitted with the Global Maritime Distress and Safety System (GMDSS). Vessels holding the certificate may be granted an extension which must not exceed 5 months until arrival at the ship's next port.

7. Certificate of Class
   Vessels built in accordance with the Lloyd's Register Rules, and which are seen to be maintained, as found by surveys to meet the standards of the society, will remain 'in class'. In the event of structural change this may be in the form of an 'Interim Certificate of Class'.

8. Tonnage certificates
   In accordance with the Merchant Shipping Tonnage Regulations (1982), British vessels will be measured by a Classification Society and issued with a British Tonnage Certificate. Other authorities (Panama and Suez) have their own tonnage measurement systems for setting costs for use of respective canals.

9. Derat or Derat Exemption, Certificate (validity 6 months)

If the Master of a ship cannot produce a Derat Certificate on arrival in a port the proper officer shall inspect the ship's condition with regard to rodents on board. If the inspecting officer thinks that the ship is free of rats he should then issue a Deratting Exemption Certificate. In the event the officer is not satisfied he should order the ship to be deratted and subsequently issue a Derat Certificate.

10. Marpol certificates

Oil Pollution Insurance Certificate (OPIC)
UK Oil Pollution Prevention Certificate (UKOPP)
International Oil Pollution Prevention Certificate (IOPP)
International Pollution Prevention Certificate for the carriage of Noxious Liquid Substances in Bulk (INLS)

11. Document of Compliance (DOC) issued under regulation 1/12 of SOLAS

This is a Document of Compliance, which for the purpose of Port State Control is treated as a certificate and issued by the MCA on completion of a successful audit to the shipping company. A certified copy being carried by the ship together with the Safety Management Certificate (SMC) issued to an individual ship must be available for inspection or the ship may be detained under Port State Control. The DOC will be specific to ship types and will be valid for 5 years but subject to annual verification $+/-3$ months.

12. Safety Management Certificate (SMC)

The Safety Management Certificate is issued to individual ships following an on-board audit of the Safety Management System aboard the vessel. This certificate will be valid for a period of 5 years and will be subject to one verification, between the second and third anniversaries. (More frequent audits may be deemed necessary and the certificate remains in force with this proviso.)

More detail on the ISM code can be found on pages 52–55.

Certificate validation and issuing authorities

It is a common practice for ships' Masters to maintain a certificate file containing relevant dates of issue, dates of expiry and the issuing authority. Understandably over a period of time Masters will become familiar with the different organization which carry out surveys and issue respective certificates. Regarding this the following background is provided on the more widely used supporting elements associated with the Mercantile Marine.

The majority of certificates are generally issued after survey and it would also be remiss not to include details of these shipboard inspections. However, with changing legislation, aspects of surveys cannot be expected to remain static and higher standards will generate changes to survey content in the future.

Classification of ships

The principal maritime nations have established their own Classification Societies, e.g. UK – Lloyd's Register of Shipping, France – Bureau Veritas, Norway – Det Norske Veritas, USA – American Bureau of Shipping, etc.

The function of the societies is to provide classification to ships based on published rules and regulations concerned with the strength and adequate provision of equipment to such vessels. Classification is not compulsory (in the UK), but an owner with an unclassified vessel would be required to satisfy the countries' regulatory bodies that the ship complied with the required standards of strength and safety to qualify for relevant operating certificates, such as Loadline or Safety Construction.

The use of a Classification Society gives the ship owner an assurance that his ship is 'well found' and without classification, such bodies as underwriters, bankers, charterers, shippers etc., might be reluctant to carry out shipping business activities. An unclassed vessel is rare today and the majority of ships are classed with one of over 45 Classification Societies.

The classification of a vessel is subject to inspection (surveys) from the time of building right through its active life, in order to stay in class. This ensures that the ship retains an expected standard whilst engaged in general trading operations.

Lloyd's Register (LR)

The LR Society acts on behalf of governments with regard to national and international requirements for inspecting and surveying vessels. Their surveyors would approve repairs to a ship's hull or to equipment on board damaged vessels, to ensure that the ship is maintained in class. Additionally, surveyors would be employed on new building or inspecting existing ships seeking classification.

Lloyd's Rules provide the following list of surveys carried out by the LR Society: New construction survey, classification survey, survey for repair/alterations, annual survey, docking survey, intermediate survey, special survey, continuous survey, statutory survey, as well as surveys for national and international shipping requirements.

Survey details

See Merchant Shipping Regulations (Cargo Ship Construction and Survey) 1981/84 SIs 572/84, 1217/84 and 1219/84.

Classification for existing ships

When classification is required for an existing ship, which has not been built under the supervision of the society's surveyors, the ship's plans and vessel's arrangement must be submitted to the society for approval. These plans should include details of the scantlings and the manufacturer and/or evidence of testing of materials used in the vessel's construction.

In the event that such plans are not available the Lloyd's surveyor should be allowed to obtain relevant information directly from the ship. The special survey for hull requirements would be carried out on ships up to 20 years old. Vessels over 20 years of age must also comply with the hull requirements for the special survey and oil tankers must comply with the additional stipulated requirements of the rules and regulations.

Surveyors will normally assess standards of workmanship, general arrangements, equipment on board and the scantlings of the vessel, all of which is submitted for approval. The date of classification is from the date of the special survey.

Classification for new tonnage

Vessels built in accordance with the Lloyd's Register Rules or equivalent standards are assigned a Certificate of Class, recorded in the Register Book. The vessel continues to remain in class as long as she is maintained according to the standards set by the rules.

New ships are built under the societies' Special Survey, which requires the presence of the surveyor from the time of commencing work until the ship is completed. Once classed the vessel will be entitled to the distinguishing mark of a 'Maltese Cross' inserted in the register before the notation of class. This emblem indicates that the ship has been constructed under the societies' Special Survey Rules. The date of completion of the survey under construction is normally taken as the date of build in the Register Book.

Types of survey

Annual surveys

The annual survey is applicable to all ships (+/ − 3 months). The exception could be a tanker which has undergone an intermediate survey within 3 months of its annual date. The annual survey very much parallels a loadline survey and is usually conducted at the same time.

In addition to an inspection of the general condition of the vessel, the annual survey should include inspection of the watertight integrity of the weather deck, especially around hatches, bow and stern doors, skylights, deck houses, scuttles and deadlights, etc.

The surveyor should also inspect scuppers, freeing ports, sanitary discharges, guard rails, gangways, life-lines, and bulwarks. Specific items of attention should be any inert gas plant aboard the vessel, together with venting systems, ballast tanks, fire protection systems, machinery and boilers. Vessels that are assigned 'UMS' notation will also have bilge level and alarm systems checked.

The steering gear and auxiliary steering method should also be inspected, as well as the freeboard marks on the ship's side. (A tanker vessel should have additional, specific requirements for an annual survey.)

Docking surveys

A docking survey is carried out twice in 5 years if the vessel is less than 15 years old. The maximum time between surveys should not exceed 3 years. (NB. One of these surveys may be replaced by an in water survey, but two successive in water surveys are not permitted.)

On vessels 15 years or older docking surveys are conducted every 2 years or every 30 months if high resistant paint is applied to the hull.

The survey pays particular attention to the structure and the possibility of corrosion, damage or general deterioration especially about the shell plate of the hull, the stern frame and rudder assembly. The propeller and sterntube area are also inspected, together with all other openings and valve arrangements which could be exposed to the sea.

In water survey

The regulations allow an in water survey to be conducted in lieu of a docking survey provided that the ship is less than 15 years old and is not a passenger vessel. The in water survey requires the same information as a docking survey, and to this end divers and/or video displays are employed for further inspection of the hull.

Intermediate survey

The intermediate survey is carried out on tanker vessels which are 10 years of age or more. This survey should be carried out 6 months before or after the halfway date of the period of validity of survey. The objective of the survey is to ensure that the tanker remains in a capable condition to allow the vessel to continue trading. It should normally include inspection of all items in the annual survey together with a more intensive hull inspection.

Anchors should be lowered and raised and at least two cargo tanks should be inspected internally. Special attention should also be paid to clearances on rudder bearings, propeller shaft and seals together with any testing required of weather deck piping arrangements. The boiler survey should also be noted and confirmed within this inspection.

Special survey

A special survey is conducted every 4 years but may be extended by 1 year on the proviso that the vessel is submitted for a 'general examination' to assess the overall condition of the ship and whether it is capable of operating through the extended period.

The special survey includes all the items considered for inspection in the annual and docking surveys and would cause the vessel to be dry docked.

Each special survey is more detailed and thorough than the last and becomes ever more so with the increasing age of the vessel. Very old ships could expect X-ray and ultrasonic tests on welds or sections of shell plate. Drill testing for assessing plate thickness could also become the order of the day.

The ship's tanking system could expect to be cleaned and examined internally with particular attention being given to structural tanks under boiler spaces. (NB. Fresh water, bunker, and lubricating oil tanks are usually exempt from the earlier surveys.)

Special surveys should also take account of cargo holds and deck features such as masts and rigging, anchors and cables, cable lockers, deep tanks, etc. for any signs of deterioration.

Additional certificates for specific vessels

Passenger vessels: Passenger Ship Safety Certificate

Period of validity: 1 year, may be extended only until the ship reaches her next port, but cannot be extended beyond a period of 5 months.

This certificate is issued by the MCA to passenger vessels which have been surveyed with:

- an initial survey when the vessel enters service,
- a periodic survey every 12 months,
- additional surveys as they may be required.

The initial survey

On entering service this survey should cover:

- The ship's structure, machinery, equipment, and the inside of boilers.
- Scantlings, electrical installations, radio communications inclusive of radio installations in lifeboats.
- All life saving appliances, and fire-fighting systems, navigation instruments, pilot ladders and pilot hoists, lights and shapes and sound signalling apparatus.

Periodic surveys

These include any of the above items but could be more detailed with stricter testing imposed.

Additional surveys

These surveys are carried out when any changes have been made to the ship.

Court of Survey

In the event that a surveyor refuses to provide a declaration of survey on a passenger vessel, the Master or the owner of the vessel may appeal to the Court of Survey.

The court constitutes a judge and two assessors who will receive arguments when:

- a ship is refused a certificate because of the lights and fog signals,
- the ship has been ordered to be detained as being unsafe,
- the ship is refused the certificate.

Continuous survey

This survey is an alternative to a special survey. It is a continual testing and survey operation of all relevant parts and equipment, conducted in rotation over a 5-year period, between consecutive examinations. Although normally carried out by a Lloyd's surveyor, the Chief Engineer may be allowed to carry out the survey of specific machinery parts in the event that a designated surveyor is not available.

Periodic survey

The periodic survey is conducted to ensure that an issued certificate can remain in force and effective. It is applicable to all ships and is generally conducted halfway through the certification period (+/−6 months) for the hull and watertight integrity.

- Steam boilers surveyed at an interval not exceeding 2 years.

- Other boilers up to 8 years old surveyed at intervals not exceeding 2 years.
- Other boilers, 8 years or older, surveyed annually.
- Propeller tube shafts surveyed at intervals not exceeding 5 years.
- Other propeller shafts surveyed at intervals not exceeding 30 months.

Notice of survey

Although the responsibility of ensuring that all the respective surveys are carried out at the required times rests with the owners of the vessel, Lloyd's Register will provide information by letter or computer printout of surveys due.

Withdrawal or suspension of class

In the event that an infringement of the regulations occurs, Lloyd's may withdraw classification of the vessel.

A ship putting to sea overloaded or operating in areas other than those designated for which she is classed, may lead to automatic suspension of class.

Pollution controls

With several high profile oil tanker incidents taking place it was to be anticipated that tighter controls would follow. Examples of this can be seen with:

- The *Torrey Canyon* wreck (1967) followed by the MARPOL convention being adopted by IMO in 1973.
- The *Amoco Cadiz* (1978) – MARPOL amended 1978.
- *Exxon Valdez* (1989) – Oil Pollution Act 1990.

The MARPOL convention, annexes and subsequent amendments have provided the means of monitoring and to some extent controlling pollution of the marine environment, i.e. MARPOL (1992) was amended to make it mandatory for new tankers to be fitted with double hulls, or an IMO-approved alternative design, applicable to all tankers ordered after July 1995.

Although the legislation is in place, policing of the regulations must be upheld and can be implemented by inspections/surveys, positive monitoring methods by Marine Authority Inspectors and record keeping by ships' personnel – a key element of inspection in such areas as cargo log books, garbage record log, etc.

Tanker vessels

Oil tankers are subject to the following surveys: initial, annual, periodic, unscheduled and intermediate.

The vessel should have all tanks and cofferdams clean and in a gas-free state. In addition to the inspection of the hull and machinery condition, tankers should also have the following examined: cargo tank openings, pipeline systems for crude oil washing (COW), pump rooms, bilge stripping arrangements, pressure vacuum valves, flame arresters, and electrical installations in hazardous areas. (The inert gas (IG) system and fire-fighting appliances should be surveyed prior to the issue of the Safety Equipment Certificate and in any event the IG system should be examined at each docking period.)

If a tanker is over 10 years old, surveys are carried out under much stricter conditions and with greater detail. The survey time must also fall within 6 months of the halfway date of the certificate validity.

In the event of an occurrence between surveys which could invalidate the survey, the Master is obligated to inform the Marine Authority which monitors the vessel and also the Marine Authority of the country in whose jurisdiction the vessel is.

Liquid gas carriers

Gas carriers undergo extensive annual and special surveys, where inspected items should include specific attention to: cargo level alarm systems, cargo control elements and leakage alarms, pipelines, cargo tanks, insulation, and reliquefication plant.

Additional inspection should be made of the cargo operation log and the 'Certificate of Fitness' which all gas and chemical carriers must possess, relating to the nature of the cargo they expect to carry.

NB. Vessels built after 1 July 1986 must comply with the International Gas Code which requires a 'Certificate of Fitness' for the carriage of liquefied gases in bulk. This is a mandatory requirement under Chapter 7 of the SOLAS convention. The period of validity is 5 years and no extension is permitted. Validity would cease if the vessel is in contravention of survey or is transferred to another flag state.

Chemical tankers

Chemical tanker vessels which are built after 1 July 1986 must comply with the International Bulk Chemical Code. This is a mandatory requirement under both the MARPOL and SOLAS conventions.

Surveys are conducted in accordance with the mandatory Safety Certificate requirements. A Certificate of Fitness for the carriage of dangerous chemicals in bulk should be issued by the MCA. The period of validity is 5 years, with no extension permitted.

Cargo record book

It is a requirement that all product tankers and chemical carriers or other vessels which are designed to carry noxious liquid substances (NLS) must carry and maintain a cargo record book. This record must be retained on board the vessel for a minimum period of 3 years from the date of last entry.

The record must show any movement of NLS and the relevant tank(s) affected. Records would also include detail of tank cleaning operations, removal of residues, disposal of residues to reception facilities, any discharge into the sea, ballasting of cargo tanks, or accidental or exceptional discharge.

Every entry into the record book must be signed by the officer in charge of the operation and each page of the record must also be signed by the Master.

NB. It is a requirement that these specific types of vessels which are employed in the transportation of noxious liquid substances should conduct their operations with due regard to a Procedures and Arrangements (P&A) Manual. This manual should be inspected at survey along with other documentation by way of certificates and the cargo record book.

Panama Canal Tonnage Certificate

This certificate is issued by the MCA or a Classification Society following survey, for vessels wishing to make a transit of the Panama Canal.

The requirement is necessary under the Panama Canal Authorities Regulations. A vessel arriving at the canal would have the tonnage calculations checked and the certificate would be exchanged for one issued by the Panama Canal Authority.

The costs of a canal transit are based on the net tonnage as shown on the certificate.

Suez Canal Special Tonnage Certificate

This certificate is a requirement for vessels wishing to pass through the Suez Canal. It is issued by the MCA or Classification Society after a measurement survey.

The survey is conducted in accordance with Canal Authority Regulations, as described by the 'Tonnage Measurement of Ships – Instructions to Surveyors' (an HMSO publication).

The obtained measurement is based on the 'Moorsom System' which provides both gross and net tonnage figures.

The costs of a canal transit are based on the net tonnage figure on the certificate.

NB. Under the new measurement system, gross and net tonnage figures are ascertained by formulae and the obtained values no longer employ the word 'ton' (there are no physical units to express gross or net tonnage), e.g. a vessel would now carry a gross tonnage (GT) of 32 568; however, this would no longer be described as 32 568 gross tons.

Examples of a Cargo Ship Safety Equipment Certificate and a Record of Inspection – Cargo Ship document can be found in Figures 1.1 and 1.2.

Master's additional handover considerations

Additionally, the Master being relieved would also be expected to hand over the following items:

- A crew list together with the discharge books (usually kept in the Master's safe). In some cases, Masters may hold the Certificates of Competency of their serving officers; however, this is not the usual policy of every company and may be adopted to suit the trade of the vessel and respective inspecting officers.
- The ship's accounts with any monies and petty cash.
- The official publications (normally kept on the bridge or in the chart room).
- The medical log book and an inventory of any drugs kept on board, i.e. morphine.
- Although the Chief Officer's main priorities are the ship's stability and cargo, most Masters expect to be kept informed of changing conditions in these two areas. Subsequently, a prudent Master would also have copies of the ships stability data or have access to it, and would normally be given a copy of the current cargo plan on completion of loading.
- The ship's keys, specifically safe keys and compass magnet/locker keys.

NB. It is normal practice to expect the Chief Officer to hold the 'Register of the Ship's Lifting Appliances and Cargo Handling Gear'. This contains separate files of rigging equipment certificates, anchors and cable certificates, etc.

Similarly, the life saving appliances documentation, such as liferaft certificates (valid 12 months but can be extended by 5 months), would also be retained by the Chief Officer and/or the Master.

Special note: gas carriers and chemical tankers

These vessels would also be required to produce a Certificate of Fitness.

The Certificate of Fitness is applicable to all tanker vessels (no limits on tonnage or size). It is issued by

Name of Ship

CARGO SHIP SAFETY EQUIPMENT CERTIFICATE

This certificate shall be supplemented by a Record of Equipment (Form E)

Issued under the provisions of the
International Convention For The Safety of Life At Sea, 1974,
as modified by the Protocol of 1978 relating thereto,
under the authority of the Government of the
United Kingdom of Great Britain and Northern Ireland
by the Marine Safety Agency
an Executive Agency of the Department of Transport

PARTICULARS OF SHIP

Name of Ship

Distinctive Number

Port of Registry

Gross Tonnage

Deadweight of Ship (metric tons)

Length of Ship (reg III/3.10)

IMO Number

Type of Ship

Date on which keel was laid or ship was at a similar stage of construction or, where applicable, date on which work for a conversion or an alteration or modification of a major character was commenced

THIS IS TO CERTIFY

1 *That the ship has been surveyed in accordance with the requirements of regulation 1/8 of the Convention, as modified by the 1978 Protocol*

2 *That the survey showed that:*

2.1 *the ship complied with the requirements of the Convention as regards fire safety systems and appliances and fire control plans;*

2.2 *the life-saving appliances and the equipment of the lifeboats, liferafts and rescue boats were provided in accordance with the requirements of the Convention;*

2.3 *the ship was provided with a line-throwing appliance and radio installations used in life-saving appliances in accordance with the requirements of the Convention;*

2.4 *the ship complied with the requirements of the Convention as regards shipborne navigational equipment, means of embarkation for pilots and nautical publications;*

Name of Ship

2.5 *the ship was provided with lights, shapes and means of making sound signals and distress signals in accordance with the requirements of the Convention and the International Regulations for Prevention of Collisions at Sea in force,*

2.6 *in all other respects the ship complied with the relevant requirements of the Convention.*

3 *That the ship operates in accordance with the regulation ▪ III/26.1.1.1 within the limits of the trade area*

4 *That in implementing regulation 1/6(b) the Government of the United Kingdom and Northern Ireland has instituted mandatory annual surveys.*

5 *That an Exemption Certificate has been issued*

This certificate is valid until

Issued at

Date of issue   *(Place of issue of certificate)*

Signed

(Signature of authorised official issuing the certificate)

Name

*Delete as appropriate

Figure 1.1

MSF 1102/ REV 0297

Name of Ship

RECORD OF INSPECTION-CARGO SHIP

mca
Maritime and Coastguard Agency

This record must be kept on board and be available for inspection at all times

1 PARTICULARS OF SHIP

Name of Ship	
Distinctive Number	*Port of Registry*
Gross Tonnage	*Deadweight of Ship*
Length of Ship	*IMO Number*
UK Class of Ship	*Type of Ship*

Date on which keel was laid or ship was at a similar stage of construction or, where applicable, date on which work for a conversion or an alteration or modification of a major character was commenced

Classification Society

Date of expiry of Safety Equipment Certificate

2 NAME AND ADDRESS OF OWNER / OPERATOR

Name	*Company*
Address	*Telephone*
	Fax
Postcode	*Country*

3 NUMBER OF SAFETY APPLIANCES

The Life Saving Appliances provide for a total of [   ] *persons*

4 SIGNATURE

The equipment described in the following sections of this document was inspected on and found to be in accordance with the requirements of the Merchant Shipping Acts and associated legislation.

Signed _____ *Signature of Surveyor(s) conducting the Survey*

Name(s)

Port of Survey _____ *Date*

Office

Safe Ships Clean Seas   1/12   FORMERLY SUR 183

MSF 1102/ REV 0297

Name of Ship

5 LIFE SAVING APPLIANCES

Lifeboats

Boat No.	Description	Measurements			Cubic Capacity	No of Persons	Internal Buoyancy		Weight Fully Laden (tonnes)
		Length	*Breadth*	*Depth*			*Material*	*Cubic Capacity*	

Boat Davits

| Boat Number | Description | | | | | | | | *SWL/set* | |

Boat Davits Winches

| Boat Number | Description | | | | | *Date of Overhaul* | | *SWL/set* | | |

Boat Davits Falls

Boat Number	Type of Purchase	Rope or Wire	Material			Breaking Strain (Tensile Load)	Date of Reversal or Renewal (State which)
			Size (mm)	*Construction*			

Safe Ships Clean Seas   2/12   FORMERLY SUR 183

Figure 1.2

Name of Ship

MSF 1102/ REV 0297

LIFE SAVING APPLIANCES CONTINUED

Liferafts

Manufacturer and type	Persons	Liferaft Number (serial)	+ Rigid/ Inflatable	Date of Inspection	Stowage

HRU's

Liferaft Davits

Liferaft Davit Number and Position	Description (Manufacturers Name and Type)	Are they of sufficient strength to lower fully-laden rafts	SWL/set

Liferaft Davit Winches

Liferaft Davit Number and Position	Description (Manufacturers Name and Type)	Date of Overhaul	SWL

Inflatable Boats

Maker's Name and Type	Length	No. of Persons	Weight with equipment and engine(tonnes)	Stowage
Type of Engine				

+ State whether rigid or inflatable

Safe Ships Clean Seas

FORMERLY SUR 183

Name of Ship

MSF 1102/ REV 0297

Liferaft Davit Falls

Liferaft Davit Number and Position	Type of purchase	Material				+ Date of Reversal or Renewal
		Rope or Wire	Size dia (in mm)	Construction	Breaking Strain	

6 FIRE DETECTION AND ALARM SYSTEM

		Make	Type	No	Description
Detectors	Machinery Space				
	Cargo Space				
	Accommodation and Service Spaces				
Control and Indicating Units					
Manual Call Points	Machinery Space Cargo Space Accommodation				

7 FIRE EXTINGUISHING

		No	Description
Portable Extinguishers	Machinery Spaces		
	Crew Spaces		
	Other Spaces		

+ State whether falls renewed or reversed

Safe Ships Clean Seas

FORMERLY SUR 183

Figure 1.2 *(continued)*

Name of Ship

FIRE EXTINGUISHING CONTINUED

	No	Description
Fire Buckets		
Sand Boxes and Scoops		
International Shore Connection		
Fire Pumps excluding Emergency fire pump		
Emergency fire pump(s)		

		No	Description
Fixed Installations including CO2, Halon and water spray systems for RO/RO cargo spaces	Machinery Spaces		
	Cargo Spaces		

Fireman's Outfits

	No	Description
Breathing Apparatus		
Safety Lamps		
Axes		
Protective Clothing		
Boots		
Gloves		
Helmets		

Safe Ships Clean Seas

Name of Ship

FIRE EXTINGUISHING CONTINUED

		No	Description
Portable foam applicators; Mobile foam appliances	Machinery Spaces		
	Other Spaces		

		No	Description
Non-Portable Extinguishers	Machinery Spaces		
	Other Spaces		
Hoses Length (with coupling)			
Plain Nozzles	Outside Machinery		Diameter of nozzle outlet =
	Machinery Spaces		Equivalent diameter of nozzle outlet =
Dual Purpose Nozzles	Other Spaces		Equivalent diameter of nozzle outlet =

Safe Ships Clean Seas

Figure 1.2   *(continued)*

Name of Ship — **MSF 1102/ REV 0297**

OTHER LIFE SAVING APPLIANCES

	Manufacturer and Type	Persons	Number	Stowage
Lifejackets			32 kgs or over	
			under 32 kgs	
Buoyant Apparatus				
Immersion Suits				

Number of Donning Instructions posted

	Type	Number without fittings	Number with light signals	Number with lines	Number with smoke signals	Number with smoke and light signals	Number of QR chutes
Lifebuoys					Date of expiry	Date of expiry	
Lifebuoy Stowage/Location							

Is a table of life-saving signals available on the ship's bridge?

10 NAVIGATIONAL EQUIPMENT

Equipment	Manufacturer	Type
RADAR (1) & (2)		
ARPA		
Gyro Compass		

Safe Ships Clean Seas — FORMERLY SUR 183

MSF 1102/ REV 0297

8 EMERGENCY CONTROLS

Closing Devices	
Remote stops item/location	
Other distant controls including Oil Fuel and Lubricating Oil tank outlet valves	
Location of controls for sea inlets discharges and bilge injection	

9 EMERGENCY ELECTRIC POWER

Source of Power including rating or capacity	
If generator, means of starting	
Services supplied	

Safe Ships Clean Seas — FORMERLY SUR 183

Figure 1.2 (*continued*)

Name of Ship

MSF 1102/ REV 0297

NAVIGATIONAL EQUIPMENT CONTINUED

Equipment	Manufacturer and Type	Date last adjusted (if appropriate)
Standard Compass		
Steering/Spare Compass		
Echosounder		
Direction Finder		
Homing Device		
Speed and Distance Measuring Equipment		
Rate of Turn Indicator		
Indicators:- (i) Rudder Angle (ii) Propeller Rate of Revolution and Direction of Thrust (iii) Propeller Pitch and Mode (if applicable)		

11 NAVIGATION LIGHTS

Lantern	Primary Manufacturer	Type and Serial No	Lamp (Wattage)	Alternative Manufacturer	Type and Serial No	Lamp (Wattage)
Mast Head						
Mast Head						
Mast Head						
Mast Fore						
Port						
Starboard						
Stern						
Anchor						

Name of Ship

MSF 1102/ REV 0297

Lantern	Primary Manufacturers	Type and Serial No	Lamp (Wattage)	Alternative Manufacturer	Type and Serial No	Lamp (Wattage)
Anchor						
Not Under Command						
Not Under Command						
Yellow Towing Light						

Positioning of Side Lights

	Primary Starboard	Primary Port	Alternative Starboard	Alternative Port
Breadth of Chocks				
Length of Screen				

Spares

Spare Electric Bulbs	Port
	Starboard
	NUC
Spare Slides	Port
	Starboard
	NUC

Signalling Lantern	Maker's name and/or marks		Number provided

12 SOUND SIGNALS, SHAPES AND ADDITIONAL LANTERNS

Diameter and Position of Bell	Type of Whistle(s)	Gong	Number of NUC Shapes	Black Diamond	Additional Lanterns/Shapes

Figure 1.2   (continued)

Name of Ship **MSF 1102/** REV 0297

13 ROCKETS AND SIGNALS

		Manufacturer's Name and Description	Date of Manufacture	Date of Expiry
Line Throwing Appliance			Rockets Cartridges	
Ship's Distress Signals	*Parachute*			
	Red Star			
Lifeboat Distress Signals	*Parachute Hand Flares*			
	Buoyant Smoke			
Means provided for emergency signals				

14 MISCELLANEOUS

Damage Control Plan Exhibited		
Fire Control Plans Inside and Outside Deckhouse		
Pilot Ladder-details of equipment		
Bulwark Ladder - if provided		
Accommodation Ladder-details if point of access/sea level exceeds 9 metres		
Pilot hoist - if provided & maintenance log book		
Particulars of Stability Information		
EPIRBs, SARTs, Portable Radios etc.	*Type*	*Serial No*

Safe Ships Clean Seas FORMERLY SUR 183

Name of Ship **MSF 1102/** REV 0297

15 EXEMPTIONS AND EQUIVALENTS

Regulation/IMO Circular	*Appliance*	*Conditions*

16 SUPPLEMENTARY INFORMATION

Port and Date	*Particulars of/or subsequent inspections alterations or renewals*	*Signed*

Surveyor Notes and Comments

An Executive Agency of THE DEPARTMENT OF TRANSPORT

Safe Ships Clean Seas FORMERLY SUR 183

Figure 1.2 (*continued*)

the MCA following an annual survey. Annual surveys conducted abroad are usually carried out by the Classification Society. The period of validity is 5 years with *no* extension permitted.

Introduction to registration

Amongst the Master's handover documents is the ship's Certificate of Registry. The topic of registration is wide and covers several issues which the serving Master should be aware of. The following coverage will hopefully provide guidance not only to the benefits of British registry, but also procedures required in the event of lost certificates or when applying for a certificate for a new vessel.

Registry

It is *not* a legal requirement to register a British ship, but if registered it serves to provide evidence as to ownership. Once registered the vessel is entitled to the protection of the flag state and allows the country means to apply and enforce the laws of that country in respect of the ship.

A shipowner would find great difficulty in continuing to trade because of the numerous checks that go with registration. Without being registered, the vessel would be unlikely to obtain insurance with all the pitfalls that would entail.

Advantages of registration

- Protection from the flag state.
- British ownership carries a historical reputation for competence and integrity.
- As the registered owner, it would be possible to raise a mortgage on the vessel.
- Once the vessel is a 'known' registered vessel it is easier to obtain clearance.
- It provides *prima facia* evidence of title to ownership.
- It remains proof of the Master's authority.

A British ship (definition)

A British ship is defined under the Merchant Shipping Act 1988 as one:

(a) Registered in the UK (England, Scotland, Wales or Northern Ireland).
(b) Registered in any one of the Crown dependencies (Isle of Man, Channel Islands).
(c) Registered in any of the dependent territories (Gibraltar, Bermuda, British Virgin Islands, St Helena, etc.).

A UK ship (definition)

A UK ship is defined as one which is:

(a) Registered in the UK.
(b) Not registered under the law of any country, but which is wholly owned by persons each of whom is either a British citizen or a corporate body (i.e. a company) established under the law of a part of the United Kingdom with its principal place of business in a part of the UK.

A British ship: ownership

Persons who qualify to be a British ship owner fall into the following categories:

1. British citizens.
2. Citizens from the British dependent territories.
3. British overseas citizens.
4. British subjects under the Nationality Act 1981.
5. British nationals (overseas) who came under the Hong Kong, British Nationality Order of 1986.
6. Corporate bodies in the UK or in a relevant overseas territory who have their principal place of business in the UK or in that territory.
7. Citizens of the Republic of Ireland.

If the person does not fall into one of the above categories, they can still be a part owner, provided the majority share is owned by a qualified person.

Entitlement to registration

If a person falls into any of the above categories they are entitled to register the ship in the UK, provided that:

(a) they are resident in the UK or
(b) that if they are not resident, they appoint a 'representative person'.

A representative person is defined as that person appointed by the shipowner and is either resident in the UK or if a company has its principal place of business in the UK.

Shares in a ship

There are 64 shares in a ship which can be owned in full or in part. Only whole shares may be traded, but each share may be owned by up to five 5 persons. It is the usual practice for a company to register as the owner of a share or shares. The person or company owning more than 33 shares is the major shareholder.

Place of registration

Many ports in the UK have an official with the capability and approval to register ships. He will be appointed by the Commissioners for Customs & Excise.

Certificate of Registry (contents)

When the ship is first registered the following details will appear on the certificate:

(a) Name of the ship.
(b) The name of the port in which she belongs.
(c) Ship's particulars: tonnage, type, engines, dimensions, builders, and the Master's name with his certificate number.
(d) Details of origin.
(e) Name and address of owners and detail of shared ownership.

Following the initial registration, any changes made to the ship must be reported and endorsed on the certificate by the Registrar General of Shipping, consular officer or an officer of the court. Such changes include:

(a) Changes in ownership.
(b) Details of mortgages and the discharge of same.
(c) Alterations to the ship.
(d) Changes from one port to another port.
(e) Any change of name to the vessel.

Certain changes may necessitate the ship being re-registered. This is known as 'registry anew'.

Application to register a ship

Application to register the vessel must be made in writing by the owner.

Proof of eligibility to own a British ship must be shown, and where the ship was built (or is being built) must be indicated.

Any details of shared ownership must be disclosed together with an undertaking to cancel any other registration which may be in force outside the UK.

NB. If the ship is a new ship, a 'Builder's Certificate' must be produced. If the ship is second-hand, a 'Bill of Sale' would need to be evident with supporting documentation.

Registration overseas

Application to register the vessel outside the UK can be made through a proper officer (e.g. consular officer) for a *Provisional Certificate of Registry*, a copy of which is dispatched to Cardiff Marine Records Office, and is valid for 3 months or until the ship arrives in a port with a Registrar.

Examples of an Application to Register a British Ship and a Declaration of Eligibility can be found in Figures 1.3 and 1.4.

Provisional Certificate of Registry

Applicants for a Provisional Certificate of Registry must:

(a) Produce the existing (foreign) certificate to the proper officer.
(b) Make a declaration of ownership to include details of shared ownership.
(c) Request an allocation of signal letters for the vessel.

Prior to sailing the Master must ensure that the vessel in all ways complies with both the Merchant Shipping Act and the regulations affecting registration.

NB. The ship while on passage under a Provisional Certificate of Registry would retain her former 'official number' until she is re-surveyed and re-marked and a new 'carving note' is issued.

Lost Certificate of Registry

In the event that the original certificate is misplaced, the Registrar General of Shipping will issue a new one. If the registration document is lost abroad then the Master must apply to the appropriate official for the issue of a Provisional Certificate of Registry. This provisional certificate must be surrendered within 10 days of the ship's arrival in the UK and a new certificate obtained to replace it.

Total loss of the ship

If the vessel becomes a total constructive loss, or if she is captured by an enemy of the state, or broken up, the Master must, if possible, hand the Certificate of Registry to the Registrar, or other appropriate person.

Commercial use of the Certificate of Registry

When a shipowner is looking for finance to support the buying or building of a new ship he will without doubt

Department of Transport
Merchant Shipping Act 1995

APPLICATION
TO
REGISTER A BRITISH SHIP

READ THE GUIDANCE LEAFLET' AND THE FOLLOWING NOTES BEFORE COMPLETING THIS FORM

Warning: the ship is not registered until a Certificate of Registry has been issued.
Please write in black ink using B_OCK CAPITALS, and tick boxes where appropriate.
- Sections 1 and 4 **must** be completed in all cases.
- Sections 2 **must** be completed if the ship has been registered before in any way or in any country.
- Sections 3 **must** be completed if this is an application for a fishing vessel.

SECTION 1: DETAILS OF THE SHIP

IS THIS AN APPLICATION TO REGISTER A FISHING VESSEL?
IF 'YES' SEE SECTION 3 BEFORE COMPLETING THIS FORM   Yes ☐   No ☐

PROPOSED NAME OF SHIP

(several names should be
entered in order of preference)

PORT OF CHOICE:

RADIO CALL SIGN (if known)

IMO/HIN* No (if known)
*DELETE AS APPROPRIATE

MAKE AND POWER OF
ENGINE   kw

APPROXIMATE LENGTH
(state metres or feet & tenths)

YEAR OF BUILD

TYPE OF SHIP
(do not complete for fishing vessels)

CONSTRUCTION
MATERIAL

NAME AND ADDRESS OF BUILDER

PLACE OF BUILD
(if different from above)

COUNTRY OF
BUILD

SECTION 2: PREVIOUS REGISTRATION DETAILS

NAME OF SHIP (if different from section 1)

REGISTRATION
NUMBER

PORT OF
REGISTRATION

REGISTERED
LENGTH (m)

WHERE WAS THE SHIP REGISTERED?

ABROAD   Which country?

*AN EU or EEA
COUNTRY   Which country?

IN THE UK   No & Year of registry
                (if applicable)

*EU Number

Port letters & numbers
(if applicable)

HAS THE SHIP AN OUTSTANDING REGISTERED MORTGAGE?   Yes ☐   No ☐

*EU – European Union. EEA – European Economic Area

SECTION 3: COMPLETE FOR FISHING VESSELS ONLY

NOTE: EVERY FISHING VESSEL MUST BE LICENSED TO FISH BY THE APPROPRIATE GOVERNMENT DEPARTMENT
(E.G. MAFF) BEFORE IT MAY FISH FOR PROFIT. IF YOU DO NOT HAVE A LICENCE OR LICENCE ENTITLEMENT YOU
SHOULD CONSULT YOUR LOCAL FISHERIES OFFICER BEFORE APPLYING TO REGISTER YOUR VESSEL.

WHAT TYPE OF REGISTRATION DO YOU REQUIRE?   FULL ☐   SIMPLE ☐

FULL provides proof of title and
enables mortgages to be registered)

HAS THE VESSEL PREVIOUSLY BEEN USED FOR ACTIVITIES OTHER THAN FISHING FOR PROFIT?
YES ☐   NO ☐   NOT KNOWN ☐

FOR WHICH ACTIVITY?

Supply Transport   TRA ☐   Research   RES ☐   Pleasure   PLA ☐
Fishing (sport)   ANG ☐   Other   OTH ☐

WILL THE VESSEL BE USED TO FISH IN COMMUNITY WATERS?   YES ☐   NO ☐

**HAS THE VESSEL BEEN GIVEN A DE-COMMISSIONING GRANT OR FINANCIAL ASSISTANCE FOR
IT TO REFRAIN FROM FISHING IN ANY MEMBER STATE?**   YES ☐   NO ☐

SECTION 4: DETAILS OF THE APPLICANT

FULL NAME AND ADDRESS (please include the postcode)

TEL. No.

FAX No.

I enclose the fee of   £   :

Cheques to be made payable to "Department of Transport"

If you are the permanent agent
for the owner please tick this box ☐
NOTE: All correspondence will be sent to the owner/managing owner unless
the owner requests the registry to send it to a specified person.

Signature   Date

I/we* being the owner(s) of the above ship request that all correspondence including the
certificate of registry, be sent to
my/our* registration agent/agent*   *delete as necessary

Signature of Owner(s)

☒ **YOU SHOULD NOW SEND THIS FORM TOGETHER WITH:**
- the correct fee, if you do not know the fee contact the registry on the number below
- the Declaration of Eligibility
- in the case of companies, a copy of any Certificate of Incorporation
- Builders Certificate and/or Bills of Sale, except for fishing vessels in respect of SIMPLE registration
- Ships to be registered on Parts I and IV will need a survey for tonnage

TO: REGISTRY OF SHIPPING & SEAMEN
     PO BOX 165
     CARDIFF   Phone: 01222 747333
     CF4 5FU   Fax: 01222 747877

ROS 10 11/96

Figure 1.3

**Department of the Environment
Transport and the Regions**
Merchant Shipping Act 1995

DECLARATION OF ELIGIBILITY
TO REGISTER A BRITISH SHIP
Please read these notes and refer to them as necessary when completing this form

FOR OFFICE USE ONLY

* Part I of the register provides for the registration of merchant and pleasure vessels etc.

* Part II of the register provides for the registration of fishing vessels.

* For the purpose of registration on Part I and Part II of the register, the property in a ship is divided into 64 shares.

* A share can be owned jointly by no more than 5 persons.

* When completed you should send this form, together with the appropriate fee and supporting documents (if required) to:

THE REGISTRY OF SHIPPING & SEAMEN,
PO BOX 165, CARDIFF, CF4 5FU.

A ship is entitled to be registered on **Part 1** of the register if the majority interest (ie, 33 shares) is owned by persons qualified to be owners of British ships, namely:

(a) British citizens or persons who are nationals of an EEA country other than the UK and are established (within the meaning of Article 52 of the EEC Treaty) in the UK;
(b) British Dependent Territories citizens;
(c) British Overseas citizens;
(d) Persons who under the British Nationality Order 1981 are British subjects;
(e) Persons who under the Hong Kong (British Nationality) Order 1986 are British Nationals (Overseas);
(f) Bodies corporate incorporated in an EEA country;
(g) Bodies corporate incorporated in any relevant British Possession and having their principal place of business in the UK or in any such possession;
(h) European Economic Interest Groupings being groupings formed in pursuance of Article 1 of Council Regulation (EEC) No. 2137/85 and registered in the UK.

For information on the EEA countries and the British Dependent Territories consult the Guide Booklet.

MSF 4704/REV1297
(Formerly Ros 15)

A ship is entitled to be registered on **Part II** of the register if the legal and beneficial ownership is owned by persons qualified to be owners of British fishing vessels, namely:

(a) British citizens or persons who are nationals of an EEA country other than the UK and are established (within the meaning of Article 52 of the EEC Treaty) in the UK;
(b) Bodies corporate incorporated in an EEA country with a place of business in the UK;
(c) European Economic Interest Groupings being groupings formed in pursuance of Article 1 of Council regulation (EEC) No. 2137/85 and registered in the UK;
(d) A local authority in the UK.

NOTES ON COMPLETING THIS FORM

Section 2: Legal ownership
(The numbers in brackets refer to the corresponding columns in section 2.)

(i) Please number each owner consecutively;
(ii) Companies incorporated in the UK and British Dependent Territories must enter their principal place of business, all other companies must enter their place of business in the UK;

NOTES ON COMPLETING THIS FORM (continued)

(iii) Part I applications - enter the corresponding letter from the list of qualified persons given in the first column of this page, (a-h). If not qualified enter "N". Part II applications - no need to complete.
(iv) enter number of shares held outright;
(v) enter number of shares in a particular joint shareholding, and
(vi) enter the relevant owner numbers (including your own number).

Section 3: Beneficial Ownership, and
Section 4: Operations of the vessel-
must be completed in respect of fishing vessels only;

Section 5: Representative person/managing owner
* If none of the owners making up the majority interest (for Part I applications), or the ownership (for Part II applications), is resident in the UK, a representative person **must** be appointed. (A body corporate shall be treated as resident if it has a place of business in the UK).
* A representative person is either an individual resident in the UK, or a body corporate incorporated in an EEA country and having a place of business in the UK.
* If more than 1 owner is resident in the UK, one of them must be appointed managing owner. In respect of applications for Part I, this must be one of the owners in the majority interest.

OFFICIAL USE ONLY;
(approved name if different)

SECTION 1: DETAILS OF THE SHIP (to be completed in all cases)

IS THIS AN APPLICATION TO REGISTER A FISHING VESSEL ? Yes ☐   No ☐

Name/proposed name of ship _____   Official number (if any) _____   Length (metres) _____

SECTION 2: LEGAL OWNERSHIP (the 64 shares in the ship are legally owned as follows)

* For ships to be registered on Part I of the Register, list the qualified owners first and <u>draw a line under them</u>, before entering the unqualified owners.
* For ships to be registered on Part II of the Register, all owners listed must be qualified owners.
* You MUST refer to the notes above before completing this section.

(i) Owner number	Mr, Mrs, Ms, Other	Surname/ Company Name	Forename(s)/ Registered Office	(ii) Address/ Principal Place of Business- Place of Business (include country in all cases)	(iii) Nationality/ Country of Incorporation	Status	(iv) Number of shares held outright	jointly held shares (v) No. of shares	(vi) Owner No.(s)

Continue on a separate page if necessary and tick box ☐

Figure 1.4

SECTION 3: BENEFICIAL OWNERSHIP   (complete for fishing vessels only)

IS BENEFICIAL OWNERSHIP THE SAME AS SECTION 2 ?   Yes [ ]   go straight to section 4

No [ ]   give details of Beneficial Ownership of shares held for persons not listed in section 2

No. of shares	Beneficial owner's Surname/ Company name	Forename(s)	Address/ Registered office	Principal place of business- Place of business	Nationality/ Country of incorporation	Name(s) of Legal Owner(s)

Continue on a separate page if necessary and tick box [ ]

SECTION 4: OPERATIONS OF THE VESSEL   (complete for fishing vessels only)

	Full name/ Company name	Address/ Registered Office	Principal place of business- Place of business in UK	Nationality/ Country of incorporation
The vessel is managed by:				
The vessel is directed and controlled by:				
The vessel is chartered by:				

SECTION 5: REPRESENTATIVE PERSON/MANAGING OWNER (see section 5 of notes before completion)

either: *I/we appoint the following to be the ship's Representative person

Full Name/Company name

Address/Place of business in UK

NOTE: If there is a change to the details in this section, the Registrar must be informed within 7 days of the change occurring.

or: *I/we appoint the following to be the ship's managing owner

Owner No. (from section 2) [ ]   Name in full

SECTION 6: DECLARATION (must be signed by all those named in section 2)

*I/we declare that:

★ the information given by *me/us in this form is true to the best of *my/our knowledge and belief;

★ * I am/we are the legal owner(s) of the shares as set out in section 2 of this form;

NB.   *In respect of companies and European Economic Interest Groupings, an authorised officer of each of the companies or groupings must sign this form*

STRIKE THROUGH SECTION (a) or (b) AS NECESSARY

Owner No.	Signature	Date

(a) *for applications to register on* **Part I** *of the Register:*   *I/we further declare that:*

★ the ship is not fishing for profit;

★ to the best of *my/our knowledge and belief, a majority interest in the ship is owned by persons qualified to be owners of British ships, and the ship is otherwise entitled to be registered on Part I of the Register,

★ any Nationals of an EEA country other than the UK, who are represented in the majority interest are established (within the meaning of Article 52 of the EEC Treaty) in the UK;

(b) *for applications to register on* **Part II** *of the Register:*   *I/we further declare that:*

★ to the best of *my/our knowledge and belief, the legal and beneficial ownership of the vessel is owned by persons qualified to be owners of British fishing vessels, and the vessel is otherwise entitled to be registered on Part II of the Register.

★ any Nationals of an EEA country other than the UK who are owners of the vessel, or persons described in sections 3 and 4 of this form are established (within the meaning of Article 52 of the EEC Treaty) in the UK;

★ the vessel will occasionally or exclusively be used for fishing for profit.

*Delete as necessary

Continue on a separate page if necessary and tick box [ ]

Figure 1.4 *(continued)*

attempt to raise monies through banks and other similar institutions. Any advance to assist the venture would require security for the loan and this might be in the form of the ship or shares in the ship. Details of this transaction will be included in the registration by the Registrar.

Any person who advances money is known as the 'mortgagee' but they do not become the shipowner unless the borrower, known as the 'mortgagor', defaults on the debt, in which case the mortgagee would retain the right to dispose of the property to regain his investment.

Any banker or mortgagee would look to satisfy himself that the ship is a good risk and would be keen to assure himself of certain facts:

(a) That the shipowner has personal integrity and a reputation in the shipping business.
(b) The current and future prospects of the shipping industry and the trade for which the vessel is to be employed.
(c) That the vessel is worth the money, and is 'in class' and expects to stay that way for the period of any outstanding mortgage.
(d) That the ship retains registry and is free of liens and charges.
(e) That the vessel is kept insured and nothing will infringe the insurance.

Penalties could be imposed: if the ship owner/borrower of any monies fails to make the required payments on the mortgage, or fails to keep the vessel in a seaworthy condition; or if the shipowner goes bankrupt, or if the ship is arrested.

NB. If the borrower is a company, then the Registrar of Companies would need to be informed of transactions, in addition to the Registrar General of Shipping and Seamen. This would ensure cross-checks on the mortgage, so that the vessel could not be mortgaged to another person without the former being made aware of this fact.

Introduction to the Official Log Book

During the handover procedure between the outgoing and relieving Masters, a statement relating to the change of command should be entered into the Official Log Book. This final act from the outgoing Master will in all probability be proceeded by a brief résumé of the contents and statements therein. (A similar procedure would be expected to take place between the Chief Officer and the Deck Log Book, and the Chief Engineer and the Engine Room Log Book.)

The Official Log Book is issued by the Registrar of Shipping, usually with the ship's articles and crew agreement at the commencement of the voyage. It is that record which reflects the legal standing of the vessel and carries the official record of the activities of the ship.

The ship's Official Log Book (OLB)

The keeping and maintaining of the Official Log Book (OLB) is governed by the Merchant Shipping (Official Log Books) Regulations.

There are some 45 types of entry listed by the regulations that must be entered into the OLB. The Master, may at times ask whether a particular 'item' be entered in the log. If in any doubt, the answer is 'yes', on the basis that if it doesn't belong it can always be ignored. If it is not entered then the Master may be in error.

The Master must be able to produce the log book on demand by the Registrar General of Shipping, a superintendent, or a proper officer. Failure to produce the log book on demand would be considered an offence.

Entries once made in the book should not be cancelled or amended. Any error which is subsequently noted may be corrected by a further entry being made. The OLB is admissible in court as evidence and as such it is considered one of the ship's most important documents.

If an entry is made which is defamatory, no legal proceedings can be taken against the person making the entry as he or she makes the entry from a position of privilege.

In some cases entries may be lengthy and it is permitted to use additional sheets. Where crew turnover is large (for instance, as with passenger vessels) computer printouts are also permitted.

On completion of the voyage, after the last man has been discharged from the vessel, the Master must deliver the log book to the Registrar, superintendent or proper officer within 48 hours.

The cover of the Official Log Book will contain the following information:

1. Name of ship, Port of Registry and the ship's official number.
2. Gross and registered tonnage.
3. Name of Master(s) and respective type and number of Certificate of Competency.
4. The name and address of the registered managing owner, ship's husband or manager.
5. Date and place at which the log book is opened.
6. Date and place at which the log book is closed.
7. The date and place of delivery to the superintendent or proper officer.
8. Date received by the Registrar General of Shipping and Seamen.

Example entries

1. Date and time of arrival in port and departure date and time.

2. A record of musters and drills, training with life saving or fire-fighting appliances.
3. When drills are not carried out in accordance with the regulations and the reason why the drill was not conducted.
4. Casualty reports:
   (a) loss or presumed loss, stranding, grounding, abandonment or damage to ship,
   (b) loss of life by fire or accident when aboard the ship or ship's boat,
   (c) any damage caused by the ship.
5. A record of every distress signal observed or received by the ship. (Together with an account of the action taken by the Master, if any. If no action is taken the reason for no action being taken.)
6. A record of where a seaman is left behind.
7. Record of inspections of crew/provisions/water, etc.
8. A record of tests and checks of steering gear.
9. Where three or more seamen complain about food or water.
10. When a Master thinks an officer is unfit for duty due to:
    (a) incompetence,
    (b) misconduct,
    (c) serious negligence,
    (d) failure to give assistance and information following a collision, or
    (e) any other reason
11. Records a change in rerating of a seaman.
12. When there is breach in the code of conduct.
13. When a Master thinks a crew member should be prosecuted under the the Merchant Shipping Act.
14. A record of births and deaths. In the case of a birth aboard the ship the mother's name is to be recorded. In the case of a death a full account should be made with regard to cause and circumstances. Any property of the dead seaman must also be recorded.
15. Illness or injury occurring on board the vessel.
16. A record of all draughts, freeboards and water density each time the vessel prepares to put to sea.
17. Any complaints by a seaman about the Master or other seaman.
18. A record of any seaman who is convicted by a court.
19. An account of the vessel sailing with one less Deck/Engineer Officer than regulations demand.
20. Conveyance orders.
21. Record of inspections of the anchors and cables.
22. A record of testing and inspecting pilot hoists.
23. Any reduction in the provisions and reasons why, together with the duration of reduced provisions.
24. Disciplinary action where the Code of Conduct does not apply.
25. Fines, remission of fines, appeals against fines together with any relevant details.
26. Loadline particulars.
27. The operation of water-tight doors closing/opening when the vessel is at sea.
28. Disputes regarding the pay of seamen.
29. Record of every seaman discharged, with date, time and place.
30. Record of inquiry into a seaman's death.

All entries should be comprehensive. In the event of an inquiry, Masters can expect statements entered into the Official Log Book to be examined in conjunction with other documents such as charts, engine movement records, Deck and Cargo Log Books etc.

The Oil Record Book (ORB)

It is a requirement under the Prevention of Pollution Regulations that every non-tanker vessel of 400 GT or over, and every tanker of 150 GT and over, must carry an Oil Record Book (ORB) (Part I – Machinery space operations). Tanker vessels over 150 GT must also carry an ORB (Part II – Cargo/ballast operations). It is a general recommendation that smaller vessels under the specified size should also keep similar records.

The regulations also specify that the ORB, whether it is part of the Official Log Book, or the Engine Room Log Book, or otherwise, must be in a format as set out in Schedule 2 of the regulations. It must also be retained in a place so as to be readily available for inspection for a period of up to 3 years after the last entry has been inserted (see *Seamanship Techniques*, Part 2 – non-tanker entries).

Example entries: tanker vessels

Loading of oil cargoes.
Unloading of oil cargoes.
Internal transfer of oil during the period of the voyage.
Crude oil washing operations.
Cleaning of cargo tanks.
Ballasting of cargo tanks.
Ballasting of dedicated clean ballast tanks (CBT tankers only).
Discharge of any dirty ballast.
Discharge of clean ballast contained in cargo tanks.
Discharge of water from slop tanks into the sea.
Discharge of ballast from dedicated clean ballast tanks (CBT tankers only).
Condition of oil discharge monitoring and control systems.
Any accidental and other exceptional discharge of oil from the vessel.
Additional operational procedures and any other general remarks.

Disposal of any residual or oily mixtures not otherwise stated.

Vessels engaged on specific trades should also include statements of:

The loading of ballast water.
The relocation of ballast water within the ship.
The discharge of ballast water to a reception facility.

Masters should note that authority has now been given by the Marine Authority for a Harbour Master or any other person designated by a Harbour Authority to:

1. Inspect the Oil Record Book of a vessel while it is in harbour.
2. Copy any entry from that book and require the Master to certify that the copy is a true record.

Official publications

Among the many handover certificates and related documents which the outgoing Master will pass over when being relieved are the Official Publications. These should be the current editions. Relieving officers are expected to acquaint themselves with respective items relevant to their own areas of work, i.e. Second Officer/Navigation Officer – Chart Correction Log, Current Weekly Notices to Mariners, Effective Navigation Warnings, etc.

In 1998 a change was made to what was previously known as '*M Notices*'. With a revised format these are now issued as Merchant Shipping Notices (MSNs), Marine Guidance Notes (MGNs) and Marine Information Notes (MINs). These are included in the Official Publication list required by the vessel and are expected to be kept up to date.

The latest Admiralty Weekly Notices/supplements and corrections can be obtained from a shipping office usually via the ship's agents. (Where electronic navigation chart (ENCs), are carried corrections are in compact disc format.)

The following is the current full list of Official Publications:

The International Code of Signals
The Mariner's Handbook
Nautical Almanac
Nautical Tables
Sailing Directions (relevant to the trade)
Admiralty List of Lights
Admiralty Tide Tables
Admiralty Lists of Radio Signals (8 volumes)
Tidal Stream Atlas
MSNs, MGNs, MINs

Weekly Notices to Mariners
Operating and maintenance manuals for navigational aids carried
Code of Safe Working Practice
A full set of corrected working charts

It should be noted that a well-found vessel would be expected to carry other publications relevant to the day-to-day working of the ship. However, the Master would be looking to ascertain the correctness of the Official Publications during any handover period.

Cargo handling gear (Hatches and Lifting Plant Regulations)

Although the topics of stability and cargo are generally within the domain of the ship's Chief Officer the documentation falls under the Master's overall responsibility. Under the International Safety Management (ISM) Code, implemented into UK law by the ISM Regulations, effective from 1 July 1998, the ship must operate a management system which will directly affect safety or pollution prevention issues.

As such, organizational practice must reflect a safe environment for the operation of lifting plant and cargo operations. The requirement is that the vessel will have a Register of Ship's Lifting Appliances and Cargo Handling Gear, and that this will be administered usually by the Chief Officer, and kept up to date with any changes to the lifting or handling gear.

The register will contain the certificates of test and reports of thorough examinations and surveys of lifting appliances together with any items of loose gear, e.g. blocks, shackles, hooks, bridles, chains, etc.

General practice indicates that Chief Officers maintain this register in a series of office files which can be readily accessed for ISM audit.

For further information see Code for Safe Working Practice, Chapter 7.

Figures 1.5 and 1.6 show examples of lifting gear which must be covered by the Register of Ship's Lifting Appliances and Cargo Handling Gear.

Master's duties: ship's arrival/departure to and from port

Entry inwards (UK legislation)

On arrival at a port the ship will be subject to inspection by:

(a) The medical authorities of the port.
(b) A government inspector (UK – possibly an MCA surveyor).

Figure 1.5   Rigging for 'Velle derrick' and supporting mast stays

(c) An official from the Ministry of Agriculture & Fisheries.

The inspectors and officials generally co-operate as overlap of inspection areas often occurs.

Based on the International World Health Regulations, the inspecting officer of the ship would be the Port Medical Officer or an authorized Officer for the task. He may inspect a ship on arrival or carry out an inspection aboard a vessel already inside his district.

An inspection of the ship would take place when:

(a) The Master of the arriving vessel reports an infectious disease on board.
(b) There is already a ship in the district on which the Master suspects there is an infectious disease.

The medical inspector may, on the Master's request or on the request of the Secretary of State, examine any person on board the vessel if he has reason to believe that:

(a) he or she is suffering from an infectious disease,
(b) he or she has been exposed to an infectious disease,
(c) he or she is verminous.

Following examination the officer may:

(a) detain the person ashore or on board the vessel,
(b) disinfect his/her belongings,
(c) prevent the person from going ashore or place restrictions on shore leave,
(d) require the assistance of the Master.

Additionally the Medical Officer may require all persons on board to show a vaccination certificate. Persons who cannot produce a vaccination certificate, may be detained by Customs Authorities. In some ports Customs officers have the authority to sight vaccination certificates.

The Medical Officer has the right to examine or place in quarantine any person whose vaccination certificate is found to be unsatisfactory. However, the regulations do not allow the officer to use the ship as a quarantine hospital to isolate infectious disease if such use would delay or interfere with the ship.

In the event that the officer thinks that someone on board is suffering from an infectious disease or tuberculosis he may have the person removed to hospital, or prevent that person being removed from the ship. If the person is suffering from tuberculosis and is intending

Figure 1.6 Derricks and cranes with supporting rigging aboard a general cargo vessel

to disembark from the ship, the officer should inform the medical authority in the area to which the infected person is expecting to travel.

On arrival the Master must answer all relevant questions of the inspecting officer and inform him of anything which may cause the spread of infectious disease. The Master's report should include detail of any rodents on board the vessel and any mortality amongst such rodents.

The health clearance and report can be given and received by radio and if this is the case the vessel is often permitted to proceed directly to a designated berth. Masters should bear in mind that they will have to make a health declaration. If the Master has to give a positive answer to any of the listed questions or is in doubt as to the health status of the ship or crew, then such information must be passed by radio not more than 12 hours nor less than 4 hours before arrival.

Relevant health signals

Q   My vessel is 'healthy' and I request free pratique.

QQ   I require health clearance.
ZU   My Maritime Declaration of Health has a positive answer to question(s) (indicated by use of complements tables).
ZW   I require the Port Medical Officer.
ZY   You have pratique.
ZZ   You should proceed to anchorage for health clearance at place indicated.

Clearing inward

Vessels arriving in the UK from abroad, or which are carrying uncleared imported goods, must report to HM Customs.

The report, in duplicate, should include:

1. The Master's Declaration – A report form which contains general voyage details and particulars of the ship together with a list of on-board stores at the time of arrival. The report should be signed by the Master.
2. Crew Declaration – A report form on which the Master and crew must declare certain goods to customs, e.g. tobacco, alcohol, etc. (Notes on the form will advise what specific goods need to be declared.) Additionally, any small packages, addressed packages, or other merchandise not declared elsewhere must be noted on this report form. The report is signed by the Master.
3. A Cargo Declaration – All your cargo must be declared to the Customs Authority. This is usually done by means of a 'Cargo Manifest' or other similar administration document which shows the following information:
   A maritime transport document, i.e. a bill of lading number.
   A container identification or vehicle registration number.
   The identity number, marks and quantities of packages.
   The description and gross weight and/or volume of the goods.
   The port or place of loading.
   The port or place of origin of the goods.

NB. The cargo declaration, if not made in the form of a manifest, can be made on a pro forma produced by IMO. Alternatively, if the ship is carrying a single bulk cargo a statement may be included in the Master's Declaration. Some ports are now also in a position to accept a computerized inventory of cargo parcels and if this method is employed then a statement to this fact must be made again on the Master's Declaration.

See Figure 1.7 for an example of a Master's Declaration.

HM Customs and Excise

Part A General

Master's Declaration

Complete this form in duplicate. Complete Part A below and Parts B&C overleaf. Report and clearance procedures are explained in Notice 69.

.... (Name of shipping line, agent, etc.)

☐ Arrival ☐ Departure

For Official Use
Rotation number

1. Name and description of ship		2. Port of *arrival / departure	3. Date & time of *arrival / departure
4. Nationality of ship	5. Name of master	6. *Port arrived from / Port of destination	
7. Certificate of registry (Port; date; number)		8. Name and address of ship's agent	
9.	10. Net register tons		
11. Position of the ship in the port (berth or station)			
12. Brief particulars of voyage (previous and subsequent ports of call; underline where remaining cargo will be discharged)			
13. Brief description of cargo (see + below)			
14. Number of crew (incl. master)	15. Number of passengers	16. Remarks	
Attached documents (indicate number of copies)			
17. Cargo manifest	18.		
19.	20. Passenger return	21.	
22. Crew declaration	23.		
24.	25.	26. Are light dues payable in respect of this voyage? *Yes / No	

+ You should report the following goods: (a) tobacco products, including cigars, cigarettes, hand rolling, pipe and chewing tobacco; (b) spirits, except perfumed and medical spirits; (c) cargo in bulk or stowed loose; (d) livestock and birds not shown in Part B, including those remaining on board.

Continued overleaf ⇨

For official use

A	D		
•	•	Passenger return	
•	•	OPIC	
•	•	Safety and loadline certificates	
	•	Light dues — Inwards	
		Light dues — Outwards	
		Other / remarks	

* Report accepted
* Entry outwards accepted
* Clearance granted

Date stamp

.... for Collector

C13

* *Delete as necessary*

PCU(August 1989)

Part B Stores

If stores are held at more than one location on the ship, a separate list containing the particulars required in boxes 28 and 29 must be submitted for each location. The total quantities must be shown on this form.

27. Number of places where stores are held ....
and actual locations :- ....

29. Name of article		Quantity
Tobacco	(lbs/grammes)	....
Cigarettes	(number)	....
Cigars	(number)	....
Cigarillos	(number)	....
Spirits *(including liqueurs)*		
litre bottles	(number)	....
standard bottles	(number)	....
other size bottles	(number)	....
Wine		
litre bottles	(number)	....
standard bottles	(number)	....
other size bottles	(number)	....
containers, casks etc	(number)	....

28. Goods in shops, kiosks, etc. *(e.g. cameras, watches, radios, mechanical lighters, clocks.)*	Quantity
....	....

30. Livestock and birds *(including pets)*		
Description	(number)	....
Description	(number)	....
Description	(number)	....

31. Are any firearms carried on board as stores ? *Yes / No
If 'Yes', please state number ....

32. Gaming machines on board *(number and location)*

Part C Declaration

WARNING:- THERE ARE HEAVY PENALTIES FOR MAKING FALSE DECLARATIONS

I declare that the particulars entered on this form and on any accompanying list(s) are true and complete.

Signature .... (Master) Date ....

For official use

(Arrivals only)

*a) All stores checked } Under seal
*b) The articles ticked have been checked } No. ....

.... Signature and grade

.... Station

* *Delete as necessary*

C13 Reverse(08/89)

Figure 1.7

Entering inward with passengers on board

In addition to the other forms previously discussed, if the vessel is carrying passengers the Master should complete and sign a form known as the 'Passenger Return'. This form should be given to Customs if the vessel is boarded on arrival. If Customs do not visit the vessel, it is the responsibility of the Master to deliver the form to the Customs office or other designated delivery point within 3 hours of reaching the assigned berth.

If the vessel has not been designated a berth, then the report must be delivered within 24 hours of arriving inside the port limits.

If the vessel arrives at a mid-channel anchorage during the night-time period, and no Customs officials are present, the Master can hold off the delivery of the form until 0900 hours the following morning.

The 3-hour rule may be extended by the ship's agents, provided that they inform the Customs Authority, in writing, whether the ship is carrying livestock (including birds). This advice must be delivered during normal business hours, or at least 6 hours before the ship's arrival in port.

Immigration

The Master must not allow persons to go ashore until they have been examined by the immigration officer.

Passenger and crew disembarkation lists

On entering inward the Master should also be responsible for advising the Customs officials, in advance of arrival, of:

1. All passengers intending to disembark.
2. Any crew member who is expected to be signed off from the vessel.

Advance notification of this information is necessary to allow Customs to issue clearance for persons leaving the vessel. It must also be anticipated that Customs may wish to inspect baggage and any articles being brought into the country.

Tonnage dues

It would not be unusual for the Customs authorities to collect the amount of 'tonnage dues' payable; however, this collection would be on behalf of the Port Authority. The Port Authority would issue the receipt of payment known as the 'tonnage dues slip', and this represents

evidence of payment. In either case Customs would acknowledge and stamp the report when tonnage dues have been paid.

Notice of Readiness

Once all administration is completed for the arrived vessel the Master would serve a 'Notice of Readiness' to the ship's agents or the receiver of the cargo. Generally, the discharge of the cargo and the ship's business in the port can then take place.

Hatch survey

There may be occasion when a Master will additionally require a hatch survey on arrival in port. This would occur in the event of damage to cargo, especially perishable goods. In such cases it would be usual to 'note protest' and inform the consignees of the respective parcels of damaged cargo.

Discharge of cargo certificate

Customs officers may decide to attend the vessel during any stage of the discharge where they consider the occasion necessary. Once all cargo has been discharged, the Master would lodge certificate C-1321, at the relevant Customs house (UK requirements).

Summary of documentation: entering inward

The Master would need to provide various ship's certificates and documentation as per the requirements of the country of arrival. Such documentation may include:

1. The Master's Declaration (UK Form C13).
2. The Crew Declaration (UK Form C142).
3. The Cargo Declaration (Cargo Manifest).
4. Maritime Declaration of Health (may be obtained by radio).
5. Passenger and Crew Disembarkation Lists.
6. RBD 1 (in the event of any birth or death during the voyage).
7. Tonnage dues slip.
8. Discharge books/CDC/or passports for the purpose of immigration officials.
9. A copy of the 'Document of Compliance' and the ship's 'Safety Management Certificate' may also be required together with the qualifying Certificates of Competency of relevant officers. (Requirements to meet the STCW '95 convention may have to be

identified, to prevent the vessel being detained by the maritime authorities of the country of arrival.)

10. In support of any of the above, the Master would be expected to produce: the Certificate of Registry, Safety Equipment Certificate, Loadline Certificate, Safety Radio Certificate, Safety Construction Certificate, Derat Exemption Certificate, Oil Pollution Insurance Certificate.

In addition to the documentation requirements, Masters would need to cause the usual safe navigation procedures and deck activities to be ongoing to bring the vessel into port limits, e.g. pilots, tugs, berthing details, etc.

Additionally, the stores should have been noted for the purpose of the declaration, and the 'bonded stores' closed ready for sealing by Customs officers.

See Figure 1.8 for an example of a Declaration of Goods by Ships' Crews Arriving From Outside the EC.

Entry outwards

Prior to a vessel commencing to load outward cargo she must 'enter outwards'. The process is carried out by the Master or the ship's agents on the same form as the Master's Declaration. This form should indicate the number of crew, the number of passengers on board and the list of stores intended to be carried.

The document must be signed by the Master or the ship's agent.

The ship's stores could be scrutinized by Customs officials and the agent may use a specific form for listing stores (UK – See Form X.S.17). This form must be signed by the Master or owner and is the document that is used to check the actual amounts of stores being loaded.

Once stores are loaded on board the vessel, the shipper will obtain the Master's signature on a 'shipping bill'. This would then be used to notify Customs so as to allow the checks to be made. The Customs officials would then seal the stores locker, known as the 'bond' – the seals on the bond can only be broken once the vessel is outside the 12-mile limit.

Clearance outwards

The Master or the ship's agent must report to the Customs house, and pay the 'Light Dues'. The Customs official should inspect the following documents:

(a) The Certificate of Registry.
(b) Safety certificates.
(c) Inward and outward light bills receipts.
(d) The Loadline Certificate.

(e) The Master's Declaration.
(f) The Cargo Manifest.
(g) The passenger return form.
(h) Oil Pollution Insurance Certificate (OPIC).

Once the documentation is found to be satisfactory and valid, the ship is allowed to proceed to sea.

NB. It must be assumed that under the required 'Safety Certificates', and with the current ISM legislation being established, the Master would be expected to produce a copy of the Document of Compliance (DOC) and the Safety Management Certificate (SMC) as the vessel enters inwards or outwards.

Light dues

Many countries maintain the navigation lights, marks and beacons from the national purse. The British system, however, is such that the costs are levied on the shipowners in the form of 'light dues' to cover the cost of installing and maintaining navigation buoys, lights and other navigational aids.

The payments become due when vessels visit UK ports and they are collected by HM Customs. Special arrangements exist for vessels which are engaged on regular trading to the UK, and ships which are considered as a 'short-sea trader' will be limited only to a maximum of 14 charges in any one year. For deep-sea vessels, the maximum number of charges in any year is seven.

NB. A year is considered a fiscal year from 1 April.

Charges are assessed on the 'net tonnage' of the vessel as measured under the British regulations. If the vessel is measured under a foreign system, there may be an additional surcharge levied.

Exceptions to light dues

Several types of vessels are not charged 'light dues', i.e. Her Majesty's warships of the Royal Navy, ships arriving for bunkers, fishing vessels, dredgers, vessels less than 20 tonnes, racing yachts or vessels seeking refuge.

Comment

The fairness of the system is questionable because vessels that are in transit and making full use of the navigational aids pay nothing. British shipowners are clearly

Declaration of goods by ships' crews arriving from outside the EC
(Notes for completion are shown on page 2)

Warning:- There are heavy penalties for making false declarations including possible forfeiture of goods.

H M Customs and Excise

Rotation number	Sheet of	C 142(1) Customs copy
Ship's name	From	
Master	UK Agent	

Declaration. I declare that I have read the warning and notes for completion of this form and that at the time of arrival of the above ship in the United Kingdom the answers I have given are true and complete.

Particulars of goods obtained abroad or on board the ship above the Statutory Allowance (Nil declarations are required).

For official use — Goods placed under seal on board

		Toba-cco	Cigar-ettes	Cigars	Spirits	Wine	Per-fume and/or Toilet water	Other goods	Receipt number or Clearance otherwise	Toba-cco	Cigar-ettes	Cigars	Spirits	Wine	Per-fume and/or Toilet water	Other goods
1. Signature	Rank/title							Description, quantity and value								(and officer's initials)
2. Name		Grms.	Number		Litres	Litres	Qty			Grms.	Number		Litres	Litres	Qty	
1.																
2																
1.																
2.																
1.																
2.																
1.																
2.																
1.																
2.																
1.																
2																
1.																
2																
1.																
2.																
1.																
2.																

I certify that :-   1 All crew members of this ship with goods to declare have completed this form* and the forms numbered .... to .. .. attached
†2 I have no small packages in my charge other than properly listed cargo

| Totals | | | | | | | | |

Signature of Master .... (or authorised officer) .... Date ....

C 142(1)   CD 0801/1/NL (06/94)   F 0798(June 1994)

*Delete if only one Form C142 is used   †Delete 2, if appropriate (See Note 2)

Officer's signature .... Date ....

Need only be given on last form if more than one used

Notes

1. Completing the form for declaration.

Both copies of this form must be completed and handed to the Customs and Excise officer who boards the ship on arrival or, if the ship is not boarded on arrival, copy (1) must be lodged at the place designated by Customs, within 3 hours of arrival.

The declaration must be signed by the Master or authorised officer and every member of the crew with goods to declare: marks of seamen unable to sign their name should be attested by a responsible officer. If more than one form is required, the forms should be fastened together and numbered consecutively.

2. What must be declared.

Goods which may be subject to prohibitions or restrictions (see paragraph 6) and all articles, however small the quantity, obtained abroad or during a voyage whether for yourself or for some other person and whether intended for this country or not, in excess of the Statutory Allowance.

3. What must be produced.

All prohibited and restricted goods must be produced to the Customs and Excise officer at such place as the officer requires, together with any other goods he may wish to see.

4. What is chargeable.

Most articles imported are chargeable with customs and/or excise duty and Value Added Tax when certain conditions are complied with.

5. Duty-free allowances.

Members of a crew who remain on a ship during her stay in port may be allowed under certain conditions, to keep for their own use on board quantities of tobacco products, alcoholic drinks and other goods. These goods must not be landed without the permission of a Customs and Excise officer.

Members of crew paying off a ship or proceeding on leave may land the quantities shown on page 4 of this form. Goods in excess of these statutory allowances **must** be produced to a Customs and Excise Officer either on board the ship or at such place as the officer may require. The appropriate customs office should be given advance notice of any crew paying off or leaving the ship in order that they may be cleared as early as possible.

The allowances do not apply to goods brought to the UK for sale or for other commercial purposes.

6. Prohibited and restricted goods.

Among the goods subject to import prohibitions and restrictions are:-
Controlled drugs such as opium, herion, morphine, cocaine, cannabis, amphetamines, lysergide (LSD), and barbiturates.
Counterfeit currency
Firearms (including gas pistols, electric shock batons and similar weapons), ammunition and explosives (including fireworks).
Flick knives.
Indecent or obscene books, magazines, films, video cassetes and other articles.
Most animals and birds, whether alive or dead (e.g. stuffed); certain articles derived from rare species including fur skins, ivory, reptile leather and goods made from them.
Meats and poultry and most of their products (whether or not cooked), including ham, bacon, sausage, pate , eggs and milk.
Plants, bulbs, trees, fruit, potatoes and other vegetables.
Radio transmitters (walkie-talkies, Citizen's Band radios, etc.) not approved for use in the UK.

7. Ship's surplus stores and livestock.

These must be listed on the ship's report and produced to the Customs and Excise officer first visiting the ship,.

8. Statutory allowances are shown on page 4.

CD 0801/1 R/NJ(06/94)

Figure 1.8

	Declaration of goods by ships' crews arriving from outside the EC						Rotation number		Sheet of		C 142(2) Ship's copy (Retained on board)

(Notes for completion are shown on page 2)

H M Customs and Excise

Warning:- There are heavy penalties for making false declarations including possible forfeiture of goods.

Ship's name	From	Master	UK Agent

Declaration. I declare that I have read the warning and notes for completion of this form and that at the time of arrival of the above ship in the United Kingdom the answers I have given are true and complete.	Particulars of goods obtained abroad or on board the ship above the Statutory Allowance (Nil declarations are required).							For official use Goods placed under seal on board							
1. Signature 2. Name	Rank/ title	Toba-cco	Cigar-ettes	Cigars	Spirits	Wine	Per-fume and/or Toilet water	Other goods Description, quantity and value	Receipt number or Clearance otherwise	Toba-cco	Cigar-ettes	Cigars	Spirits	Wine	Per-fume and/or Toilet water	Other goods (and officer's initials)
		Grms.	Number		Litres	Litres	Qty			Grms.	Number		Litres	Litres	Qty	
1. 2.																
1. 2.																
1. 2.																
1. 2.																
1. 2.																
1. 2.																
1. 2.																
1. 2.																
1. 2.																
1. 2.																

I certify that :- 1. All crew members of this ship with goods to declare have completed this form* and the forms numbered .... to .... attached
†2 I have no small packages in my charge other than properly listed cargo

Totals						

Signature of Master (or authorised officer) .... (Need only be given on last form if more than one used)   .... Date   *Delete if only one Form C142 is used   †Delete 2, if appropriate (See Note 2)

Officer's signature ....Date

C 142(2)   CD 0801/2/NL (06/94)   F 0798(June 1994)

†Statutory landing allowances for paid crew members

The following allowances are applicable only to those members of the paid crew on arrival in the United Kingdom.

Relief from payment of duty and VAT is conditional upon the goods being produced for examination as the officer may direct.

When landing from a ship arriving from a foreign port outside the European Community*.

Tobacco products

Cigarettes 200
or cigarillos 100
or cigars 50
or smoking tobacco .... 250 grammes

Alcholic drinks

over 22% vol. .... 1 litre
or not over 22% vol; fortified or sparkling wine .... 2 litres
plus still table wine .... 2 litres
Perfume .... 60 ml
Toilet water .... 250 ml
Other goods £136

†These allowances do not apply to goods brought in for sale or other commercial purposes.

* The countries of the European Community are:

Belgium	Luxembourg
Denmark	The Netherlands
France	Portugal
Germany	The Republic of Ireland
Greece	Spain (not the Canary Islands)
Italy	The United Kingdom (not the Channel Islands)

CD 0801/2RNJ(06/94)

Figure 1.8   *(continued)*

against the system and continue to lobby the government of the day to abolish 'light dues'.

Summary of documentation: clearance outward

1. Master's Declaration (UK C13).
2. Any other clearances issued by other European Community ports for the same voyage.
3. Cargo Manifest.
4. Tonnage Certificates and Loadline Certificate.
5. Passenger Return (Departure) (UK form PAS 15).
6. Certificate of Registry.
7. Inward and outward light dues receipts.
8. Safety Equipment Certificate or any of the other ship's papers which may be requested.
9. Oil Pollution Insurance Certificate.
10. Document of Compliance (copy) and Safety Management Certificate.

When preparing for an outward bound voyage Masters should ensure that they have an adequate supply of additional documentation such as:

- RBD 1 forms, Wage Account forms and National Insurance forms (UK forms BF 19 and 19A), Income Tax Return forms (UK 46M), Freeboard Particular forms (form 13), latest MSNs, MGNs, MINs, Up-to-date navigation warnings, and the latest Weekly Notices to Mariners.
- The Master should also order pilots, tugs, wharfmen, etc., in preparation for letting go from the berth and getting underway.
- The prudent Master should also ensure that the vessel is carrying sufficient bunkers and fresh water inclusive of a reserve, to make the next port of call.

Masters: reportable incidents

Under the Merchant Shipping Act and the SOLAS regulations ship's Masters are expected to report a variety of incidents/items, etc.

Safety of Navigation SOLAS requirements are such that the Master must report

- when dangerous ice is sighted,
- when a derelict is sighted,
- When storm force winds of force 10 or above are experienced for which no report has been given,
- when subfreezing air temperatures are experienced.

In addition to the above Masters are expected to report numerous other items and topics of interest.

Accidents

Following the *Herald of Free Enterprise* accident in 1988 the Marine Accident Investigation Branch (MAIB) was established. This is an independent body which is made up of a number of professionals from the three main disciplines of nautical, engineering, and naval architecture, with various support staff. The MAIB reports directly to the Secretary of State for Transport, and bypasses the normal Marine Coastguard Agency bureaucracy.

The investigations of the MAIB are to determine the cause of accidents in order to prevent them happening again.

Inspectors of the MAIB are expected to travel to the scene of an accident and investigate such accidents when:

(a) There has been loss of life or serious injury to a person on board a vessel, or if any person is lost from the ship or from a ship's boat.
(b) A ship is lost or presumed lost, or has been abandoned or materially damaged.
(c) A ship has stranded or been involved in a collision.
(d) A ship has caused material damage.
(e) A ship has been disabled for more than 12 hours, or less, and if as a result of her disablement she needs assistance to reach port.

For an MAIB Incident Report Form, see Figure 1.9.

Legal requirements of ship's Masters

It is a legal requirement for the Master of a British ship to report any casualty to the MAIB as soon after the event as possible and in any case not more than seven days after the ship's arrival in the next port of call.

The report should contain the following information:

- The general particulars of the ship: name, official number, Port of Registry, position, next port of call, and ETA if possible.
- Details of the incident with regard to a general description and details of any injuries to personnel.
- Damage to ship, etc.

NB. An appropriate entry should also be made in the Official Log Book in the event of the death of a seaman, together with an inventory of the property of the dead crew member.

The subsequent actions, if any, by the MAIB would fall into the following categories:

MAIB
MARINE ACCIDENT INVESTIGATION BRANCH

Incident Report Form

Annex A

For Official Use
Ref
Code

- The Merchant Shipping (Accident Reporting and Investigation) Regulations require Masters, Skippers and Owners to report accidents and dangerous occurrences. They are encouraged to report hazardous incidents as well. The terms are explained in the Regulations and in the Merchant Shipping Notice on accident reporting. Briefly, they include any accident leading to death or significant injury, or to loss or abandonment of the vessel or to her suffering material damage; any stranding, collision, fire, explosion or major breakdown; any incident causing harm to any person or the environment; and any incident which might have led to injury or which hazarded the ship.

- Please read the Merchant Shipping Notice for further details and advice, or telephone MAIB on 01703 395500.

- One form should be completed for each incident.

- Please return the completed form to: Marine Accident Investigation Branch
First Floor, Carlton House,
Carlton Place,
Southampton, SO15 2DZ,
United Kingdom

- **Completing and signing this form does not constitute an admission of liability of any kind, either by the person making the report or any other person.**

- Please complete the form clearly, using black or blue ink.   Please [✓] the boxes.

Section A

	Day	Month	Year
Date of Incident			

Time of Incident (state whether UTC (GMT) or local time):

Name of vessel

Previous name (if changed in last 6 months)

If fishing vessel please state type (eg stern trawler, crabber etc)

Official Number or Fishing Number or (if non-UK) Call Sign

Name and Port of Registry or Flag of any other vessel involved

Name and address of owner or manager

Tel. No.

IRF (1/98)

DETR
ENVIRONMENT
TRANSPORT
REGIONS

Figure 1.9

Section B

Date and time of departure from last port

Location of incident (eg latitude & longitude or name of port, or other geographical reference)

Voyage from and to: From: To:

Weather and visibility at time of incident

Responsibility: was incident caused principally by persons on another vessel, or shoreside persons, or persons **not** sailing with your vessel?   Yes [ ]   No [ ]

Type of incident (please tick appropriate boxes)
Fatal injury [ ]   Non-fatal injury [ ]
Vessel lost or abandoned [ ]   Vessel damaged [ ]
Other accident or incident [ ]

Section C - Details of person(s) killed or injured

(This section should be completed if any person has been killed or injured)

Place of incident (eg engine room; galley)

How many person(s) suffered an accident which resulted in death or injuries preventing the performance of the normal full range of duties for 3 days or more after the day of the incident?

Please complete the questions in the table for each person.

Position (eg rank, rating, passenger)	Age	Injured part of body	Kind of injury	*Hours worked before incident	Duration of last off duty period	Whether on duty
						Yes No
						Yes No
						Yes No
						Yes No
						Yes No
						Yes No

* For operational staff only

If more than 6 persons suffered reportable accidents please continue on page 4.

1   2

Section F

To be completed by the ship's
Safety Officer if applicable

Signed

Signed

Name

Name

Master or
Owner's
representative

Date

Date

Section G (if applicable)

If the incident involved a reportable personal accident or was a dangerous occurrence and there is an elected Safety Representative on board the vessel, he must be shown the completed report and allowed to write in this section any comments which he may wish to make. If the injured persons are represented by different Safety Representatives, each may make additional comments if desired in the space below but in any event, they should all sign the form.

Signed

Safety Representative

Date

Name

This space may be used as an extension of Sections C, D, E and G. **Please state clearly which sections are being expanded.**

If there is insufficient space in any part of this form for your answers or comments, please use a plain sheet of paper as a continuation sheet and fasten it securely to this form. Please indicate in the box below the number of sheets used.

Number of continuation sheets

4

Section D

Please give a brief description of the sequence of events leading to the incident.

If necessary continue on page 4.

Section E

1. Please state how you think the incident happened.

2. Has any action been recommended by you as a result and if so, what?

3. Has any action been taken and if so, what?

If necessary continue on page 4.

3

Figure 1.9 (*continued*)

(a) No further action required.
(b) Additional information is sought from the owners, employer or other body.
(c) A limited investigation is required by an inspector.
(d) An inspector's inquiry is called for under the Merchant Shipping Act regulations.

Method of reporting

Masters should be advised that they are legally obliged to report the basic details of any incident. Additional information should in certain circumstances, and in the Master's own interest, not be divulged without first taking professional legal advice.

The WRE1 report

After informing the MAIB, a more detailed report should be made on the WRE1 form, available from the Receiver of Wrecks. The form is a detailed report form on a marine casualty, and the type of casualty must be listed, e.g. fire, stranding, collision.

Example: collision

The legal requirements following a collision are:

- Standby to render assistance to the other vessel.
- Exchange relevant information with the Master or officer in charge of the other vessel (see *Seamanship Techniques*, Part 2, Chapter 6, Information exchanged).
- Report the incident to the MAIB.
- Cause an entry to be made in the Official Log Book concerning the incident.

The WRE1 report should be completed at a convenient time after the incident and include such information as:

(a) Particulars of ship.
(b) Trade in which the vessel is engaged.
(c) Crew and passenger figures.
(d) Master's and Officer of the Watch details and pilot's details if applicable.
(e) Date, time and position of incident.
(f) The meteorological conditions at the time.
(g) Details of lives lost or saved.
(h) Salvage arrangements.
(i) Extent of damage incurred.
(j) Details of accident and could it have been prevented.
(k) Comments from the MAIB.
(l) Draught of ship.

(m) List of charts in use (original charts may be required in an investigation).
(n) Compass information.
(o) Detail of cargo effects on the compass.
(p) The condition of life saving appliances.
(q) The condition of fire-fighting appliances.
(r) Watertight compartment information, were they used.
(s) Condition of the vessel.
(t) Bunker situation.
(u) Cargo/ballast information (coal cargo – ventilation details).
(v) Deck cargo detail (timber cargo).
(w) Navigation aids carried and in use. Also radar range in use.
(x) Meteorological conditions at the time of loading.

The Receiver of Wrecks

Many incidents have generated what is often constituted as a 'wreck'. To deal with a defined 'wreck' under British legislation the Department of Transport appoints a district officer known as the Receiver of Wrecks. This is an official appointment, and is usually a Customs officer or Coastguard official who acts on behalf of the Department of Transport.

The powers of the officer are such that:

1. He/she may take command of any designated wreck situation and exercise direct control of all persons involved. (However, the officer would not interfere with the operational role of the Master unless requested to do so.)
2. The officer can requisition transport and the assistance of nearby vessels to resolve the incident.
3. The officer is permitted to cross property in order to gain access to the coast.
4. The officer may cause plunderers of the wreck to be apprehended.
5. In the event that a person is injured or killed while resisting arrest, the officer would not be held liable.

Wreck definition

Wreck is defined as including any 'flotsam', 'jetsam', 'lagan' or 'derelict' found in or on the shores of the sea or any tidal water.

'Flotsam' is defined as goods which are thrown overboard and are recoverable by the fact that they remain afloat.

'Jetsam' is defined as goods thrown or lost overboard from a ship, which are recoverable by reason of being

Return of Births and Deaths

RBD 1
(Rev 7/90)

**For the purpose of Regulation 5 and 6 of the Merchant Shipping
(Returns of Births and Deaths) Regulations 1979**

Important: The Consul or other Officer abroad, or the Superintendent in the United Kingdom to whom this
Return is given, should forward it without delay direct to:

> Registry of Shipping and Seamen
> Anchor House
> Cheviot Close
> Parc-Ty-Glas
> Llanishen
> CARDIFF
> CF4 5JA
> Tel (0122 747333)
> Fax (0122 747877)

Name of Ship	Official Number or, if a fishing vessel, RSS Number together with Port Letters and Number	Port of Registry (if the ship is registered outside the UK or is unregistered state also name and address of owners)

Birth

Date of Birth	Place of Birth (By Latitude and Longitude if at sea)	Forenames (if any) and Surname of Child	Sex	Father* Name, Surname and Occupation	Father* Usual Residence and Nationality

* Note If the child is illegitimate, the particulars relating to the father **must not** be given unless it is at the joint request of the mother and
the person acknowledging himself to be the father, in which case such person shall sign the Return, as well as the mother.

Death

Date of Death	Place of Death or Loss (by Latitude and Longitude if at sea)	Name and Surname of Deceased (also maiden surname of woman who has married, if known)	Sex	Date of Birth (if known) or Age

Certificate to be signed by the proper officer to whom this return is delivered

I certify that,

- This report was made to me at the date and place shown below.
- I have examined the Official Log Book containing the relevant entries.
- I have / have not held an inquiry under Section 61 of the Merchant Shipping Act 1970 (please delete as appropriate)

Signature of Officer _____

Title of Officer _____ Port _____

Date _____

Instructions to Masters[1]

A Return of Birth or Death should be delivered to a Superintendent, Consul or Shipping Master at the earliest
opportunity. The Master should produce the Return together with his Official Log Book recording the occurrence.

UK Ships (Registered and Unregistered)

The Master is required to make a Return of

- any birth of a child
- any death in the ship[2],
- the death of any person employed in the ship, where it occurs outside the UK.

Other Ships

If the ship calls at a port the Master is required to
make a Return of any birth or death of a citizen of
the UK or Colonies[3] which has occurred in the ship
during the voyage.

Instructions to Consuls, Shipping Masters and Superintendents

The Officer receiving a Return from a Master must
be satisfied that

- it is correctly completed in all particulars. Any omission or ambiguity (such as stating the cause of death simply as "missing") will lead to delay in registration.
- the person making the Return is the Master of the ship
- the vessel is a sea-going ship
- the entry in this Return and in the Official Log Book (where carried) are consistent with one another.

If a death occurred ashore, or the body was brought
ashore, it will be helpful if a copy of the Post
Mortem findings or other medical or police reports
(with a translation into English if in a foreign
language) is attached to this Return.

The Officer should also refer to his Instructions or
Notes for Guidance in order to find out whether he is
to hold an Inquiry into the cause of death.

In the event of difficulties, assistance should be
sought from RSS at the above address.

Notes

[1]Master includes every person (except a pilot)
having command or charge of any ship.

[2]In the ship includes in a ship's boat or life-raft and
being lost from a ship, a ship's boat or ship's
life-raft.

[3]Citizen of the UK and Colonies under the British
Nationality Act 1981 means a person who is a

- British Citizen,
- British Dependent Territories Citizen,
- British Overseas Citizen,
- British National (Overseas), (under the Hong Kong (British Nationality) Order 1986)

Figure 1.10

Additional particulars in respect of deceased member of the crew, if known.

Name, Relationship and
Address of Next of Kin _____

Discharge Book No _____

UK ships only. If the deceased was engaged on a 'supplementary' agreement
for non UK seamen, please state the date and place that agreement was opened.

Date _____

Place _____

Copy of entry or entries relating to death appearing in the narrative section of the Official Log Book or other official record.			Entries required by Regulations made under Section 68, Merchant Shipping Act. 1970
Date and hour of the occurrence	Place of occurrence or situation by Latitude and Longitude at sea	Date of Entry	

Mother*			Signature of Father* and/or Mother of the child
Name, Surname and Occupation	Maiden Surname and Surname at marriage if different	Usual Residence and Nationality	

When completing this form or making entries in the Log Book as to the "Cause of Death", terms such as "suicide" or "missing" should be
avoided and more specific terms such as "gunshot wound to the head" or "lost at sea believed killed or drowned" used instead.

Occupation, rank or profession	Usual Residence at time of Death	Nationality	Cause of Death	Certified correct by Ship's Doctor or other Medical Practitioner (where possible)

Certificate to be signed by the Master

I certify that,

● this Return is correct and true;

● I have recorded the death in the Official Log Book or other record of the ship;

● the extract from the Official Log Book at page 4 of this Return is a true extract.

Signature of Master _____

Full name of Master (CAPITALS) _____

Date _____

Figure 1.10   *(continued)*

washed ashore or remaining submerged in comparatively shallow water.

'Lagan' is defined as goods thrown overboard which are buoyed, to allow recovery at a later time.

'Derelict' is defined as a vessel which remains afloat, but is abandoned. (The first salvor to board a derelict, with no crew on board, can take possession and has absolute rights and control, provided it is a serious salvage attempt.)

The finder of wreck

The person who finds the wreck, or the owner of the wreck, must notify the Receiver of Wreck for that district. The finder must also deliver the wreck to the Receiver of Wreck. Failure to do this could result in a fine of up to £1000, and the forfeiture of all salvage rights.

Details of wrecks are posted in he Customs house together with their value, while the wreck itself is normally held for a period of 12 months before being disposed of.

The proceeds from any wreck (usually through public auction or other form of sale) are paid into the National Treasury.

Dangerous occurrences

Masters are obligated to report dangerous occurrences within 14 days (or within 14 days of the next arrival in port, if the ship was at sea at the time of the occurrence taking place). Additional reference should be made to SI 1994, No. 2013.

Dangerous occurrences are defined by any of the following:

(a) The fall of any person overboard.
(b) Any fire or explosion.
(c) The collapse or bursting of any pressure vessel, pipeline or valve, or the ignition of anything in a pipeline.
(d) The collapse or failure of any lifting equipment, access equipment, hatch cover, staging or Bosun's chair or any load-bearing parts.
(e) The uncontrolled release or escape of any harmful substance or agent.
(f) Any collapse of cargo, unintended movement of cargo sufficient to cause a list, or loss of cargo overboard.
(g) Any snagging of fishing gear which results in the vessel heeling to a dangerous angle.

(h) The parting of a tow rope.
(i) Any contact by a person with loose asbestos fibre, except when full protective clothing is worn.

The reporting of the above occurrences is applicable to all ships of whatever nationality, if they are in a UK port when the accident or occurrence takes place.

Failure to report

If the Master or officer in charge fails to report an accident or dangerous occurrence without reasonable cause, he shall be guilty of an offence and liable on summary conviction to a fine not exceeding level 3 on the standard scale.

Births and deaths

The Master must make a report of a birth or death aboard the ship. Additionally he should also make an entry in the Official Log Book with a statement as to the mother's name in the case of a birth, and an inventory of the seaman's belongings in the case of a crew member's death.

The report would normally be made on a standard form known as a Return of Births and Deaths (RBD) see Figure 1.10. Alternatively it could be reported in writing to a proper officer at the next UK port of call, or at the end of the voyage. The return should generally be made as soon as possible after the event and certainly within 6 months of the event.

A proper officer in this case would normally be the Registrar General of Shipping or consular officer. In the absence of the Master, the return could be made by the Registrar provided that the death does not involve a post-mortem, inquest or is subject to an inquiry.

The Registrar is responsible for informing the appropriate UK County Registrar of all births and deaths, together with the relevant details.

The obligation of notification to the next of kin is the responsibility of the Master and must be carried out within three days of a death.

It must be anticipated that in many cases, certainly following an accident, which results in the death of a seaman, an internal on-board inquiry would be ordered by the Master. The Safety Officer would be expected to be involved and the evidence relevant to the incident would be collated.

NB. Masters may find the Nautical Institute publication *The Master's Role in Collecting Evidence* a useful reference when amassing and collating evidence which might be required for a formal inquiry or in a court of law.

2 Commercial knowledge

Introduction

The role of the Master is one which must encompass change and many trades have seen the administration as a fundamental product of whatever enterprise the vessel is engaged in. More and more the Master is being required to be aware of the cost effectiveness and economic statistics of the voyage. However, this is not a new horizon, and is in some cases the return of an age-old cycle when the Master was the shipowner, and was actively engaged in securing cargoes for his ship and generating a profit.

Although the ship's Master will not necessarily deal direct with the many organizations actively engaged in commerce he will without doubt become the end user of their influence. It is to this end that background knowledge of commercial activity is desirable. Provided the venture is successful the Master would not expect to become overly involved. It is generally when operations do not proceed as expected and damage to cargo or loss of cargo is experienced that the Master's knowledge of insurance, salvage and the law tend to come to the fore.

Commercial familiarization

In order to fully appreciate commercial operations it is essential that the ship's Master becomes familiar with the common terms and phrases of the Carriage of Goods by Sea Act.

The movement of parcels of cargo around the world is achieved with a certain amount of associated documentation in order to satisfy border controls and international boundaries. Import/export procedures are expected to meet the demands of Customs and legislation of respective countries to avoid time-consuming delays.

The ship as the carrier

The shipping of goods and the conduct of trade can be traced back many centuries and the ships that continue

the practice today operate as either a common carrier or a private carrier.

The common carrier is one who holds himself ready for hire or reward to transport goods from one destination to another. The common carrier is used by any person who wishes to employ him and the customer must deliver the goods for shipment to the carrier.

The carrier is responsible for the safe delivery of the goods without any unreasonable delay. He may be assumed to be the insurer of the goods and as such would become liable for damage or loss to the same, other than by exceptional circumstances. Exceptional circumstances could include any or all of the following causes of damage or loss:

- Negligence by the consignor (poor packaging), or by enemies of the state when at war.
- Jettison of the cargo for the overall good and safety of the venture.
- Inherent vice of the goods (like the rotting of fruit), by an order of restraint from a head of state, or from what is considered 'an act of God'.

The private carrier is one who does not comply with the conditions of the common carrier, and makes a private agreement with each consignor of goods. The private carrier is bound under common law but may include certain exception clauses within the contract which could relieve him of specific liabilities. Generally the private carrier is only liable if the goods are lost or damaged due to his own negligence and in the event of loss or damage being incurred then the onus falls to the carrier to prove that it was not caused by his negligence.

This is probably the most popular type of carrier today where the owner or charterer will be the private carrier. Ships tend generally to operate under a contract of agreement, known as a charter party.

For an illustration of the cargo export/import cycle, see Figure 2.1.

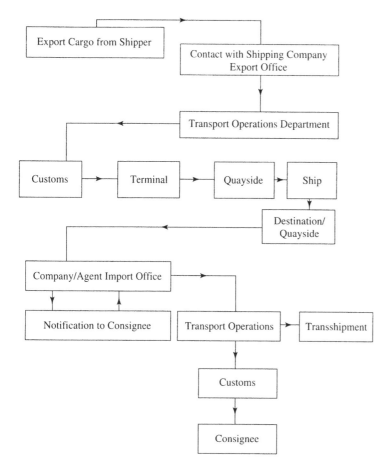

Figure 2.1 Cargo export/import cycle

Responsibilities and liability of the carrier

For additional information see the Carriage of Goods by Sea Act 1971.

Care of cargo

Any carrier must take appropriate care when loading, stowing, handling, carrying and discharging any goods carried.

Once goods have been received by the carrier into his charge, the carrier must issue a 'bill of lading' on demand which contains amongst other things:

(a) The marks necessary for identification of the goods.
(b) The quantity, weight or number of packages as stipulated by the shippers/boat note.
(c) The state of order and condition of goods as received.

The carrier may be relieved from showing the identification marks on the bill of lading if he has reason to believe that such marks will no longer be legible until the end of the shipment period. The carrier will also be relieved of showing any information which he suspects of being inaccurate or for which he has no means of checking.

NB. The bill of lading (B/L) is accepted evidence that the carrier has received the goods as described.

Shipped B/L

Once goods have been loaded aboard the vessel, the carrier must issue on demand a 'Shipped' B/L, on the proviso that the shipper surrenders any 'Received for Shipment' B/Ls.

Alternative action allows the carrier to convert the 'Received for Shipment' B/L into a 'Shipped' B/L, by endorsement of the vessel's name and the date of shipment.

NB. All bills of lading whether affecting coastal or foreign going carriers are covered by the Carriage of Goods by Sea Act (amended to the Hague/Visby Rules) when the B/L is issued in the UK, or the port of shipment is in the UK, or the B/L or other non-negotiable contract document is covered by the Hague/Visby Rules irrespective of ship nationality, carrier, shipper, or consignee. It may not cover shipment to the UK and is only good for the seaborne part of a journey which involves land or air carriage.

Delivery by the carrier

It is the responsibility of the consignee to inform the carrier or his agent, in writing, of any loss or damage incurred at the time of delivery or within 3 days of delivery of the goods.

Shipment of dangerous goods

The carrier may land, jettison, destroy or render innocuous any dangerous goods which have been shipped incorrectly marked or without the prior knowledge or consent of the carrier. If such action is demanded, the shipper would not be entitled to compensation for his goods and could be made liable for any expenses or damages incurred from the shipping of such goods.

In the event that such goods have been shipped with the full knowledge of the carrier, and they subsequently become a hazard to the well-being of the venture, then the goods can be handled in a similar way without liability, except with the consideration to 'general average' if this arises.

Seaworthiness of the carrier

For the period from the beginning of loading until the vessel commences her voyage, the carrier must exercise 'due diligence' to ensure that the ship:

(a) is retained in a seaworthy condition,
(b) is correctly manned, equipped and stored,
(c) has secure cargo compartments ready for the reception of cargo.

The burden of proof on what constitutes due diligence rests with the carrier. However, due diligence must be considered as taking all reasonable care as opposed to the absolute guarantee of what is considered 'seaworthy' as defined by common law.

Passengers: carrier liability

For additional information see Athens Convention, Carriage of Passengers and their Luggage by Sea (1974),

internationally enforced and adopted by United Kingdom legislation (1987).

Definitions

'(Passenger) carriage', is defined as that period from when the passenger embarks on the ship or the ship's tender, until disembarking from the ship or the ship's tender.

NB. It does not include time on the quayside, or in a marine terminal, or at a port/harbour installation.

'The passenger' is defined as any person who is carried for reward under a contract of carriage, inclusive of a person accompanying a vehicle.

In the event of the death of, or injury to, a passenger, or the loss or damage of luggage accompanying the passenger during the period of carriage, due to the negligence or fault of the carrier, the carrier is held liable. This is also applicable to the servants or agents of the carrier.

If injury, loss or damage are incurred directly or indirectly from collision, shipwreck, stranding, explosion, fire or defect of the vessel, then it is a presumption that the carrier is at fault or guilty of neglect.

'Luggage' is defined as any article or vehicle which is carried under a contract of carriage other than under a charter party or a bill of lading. (It does not include the carriage of live animals.)

Any loss or damage to luggage, due to the fault or neglect of the carrier, during the period of carriage renders the carrier liable. This would also include financial loss incurred due to its late delivery after the arrival of the vessel but would not include delays caused by labour disputes.

Items of high value

The carrier is only liable for loss or damage to valuable items, if such items are deposited with the carrier for safe keeping. High value items include monies, gold, jewellery, securities, etc.

Immunity of the carrier

The carrier would not be held responsible for loss or damage which results from any of the listed 'excepted perils':

1. Any neglect or default by the Master, crew or pilot or other servant of the carrier in the navigation or management of the vessel.

2. A fire, unless caused by actual fault or privity of the carrier.
3. An act of war.
4. An act of God.
5. An act by any public enemy.
6. Any omission by the shipper, owner, agent or their designated representatives.
7. A strike, stoppage, or lock-out or other labour restriction from whatever cause.
8. Riot or other civil disruption.
9. Arrest by a country's ruler or head of state.
10. Any latent defects not discovered by due diligence.
11. Saving or attempting to save life or property at sea.
12. Any wastage in bulk or weight through damage or inherent vice of the cargo.
13. Any ineffective packaging of the goods.
14. Any ineffective marking of the goods.
15. Quarantine restrictions placed on the vessel.
16. Any other cause which may arise from the actual fault or privity of the carrier or his servants (the onus of proof being placed on the party claiming the exception, to show the actual fault or neglect contributed to the loss or damage).

Reference to the Hague/Visby rules

In 1921 the International Law Association met at the Hague to establish the rights of shippers and the liabilities of the carrier for cargo being carried under a bill of lading. These rules were updated in 1968 by the Brussels Protocol to meet the needs of modern shipping practice and became known as the Hague/Visby Rules.

Not every country agreed to these rules, which the UK adopted and subsequently passed the Carriage of Goods at Sea Act 1971.

Some countries continue to use the old Hague Rules, while others use the Hamburg Rules, established in 1978 (these are considered more favourable to shippers and importing merchants and more beneficial to third world countries).

A 'charter party' is defined as a contract of agreement between a shipowner and a charterer for the purpose of hiring a vessel. There are three main types of charter party:

1. A voyage charter is used when the vessel is hired for a specified voyage or voyages. The ship must be in a seaworthy condition and hold the correct documentation/certification. She would also be cargo worthy, that is in a fit state to receive cargo at the date of commencing the charter. The owners would appoint the Master to the ship.
2. The time charter is used when the vessel is hired for a specified period of time, which may be a short- or long-term period. The time of delivery to and from the charterer is an important issue as charges are incurred from commencement to completion of the agreement. The 'time charter' does not carry the same requirement for seaworthiness as the voyage charter. The owners would appoint the Master to the ship.
3. The bare boat charter is an agreement where the charterer takes full responsibility for the vessel and supplies his own crew and conducts his own repairs. The charterers appoint the Master to the ship.

Additionally the demise charter, in the case of dry cargo ships, would be as per a bare boat charter, and in the case of tanker vessels, would be an agreement where the shipowner continues to manage the ship to the charterer's instructions under a management agreement clause.

NB. 'Seaworthiness' is defined as that condition of fitness that an ordinary, careful and prudent owner would require the vessel to have at the commencement of her voyage, having due regard to the circumstances of that voyage.

Charter party and the Master

In some cases the owners may intend the Master to conduct business on their behalf and if this is the case then it is essential that the Master is in possession of a copy of the charter party (C/P). Without a shipboard copy it would be difficult to negotiate types of cargoes, collection of freight, arrange loading and discharge or berthing arrangements. Alternatively, company policy could stipulate the order of business via the company's appointed agents.

In any event the Master should be familiar with the various clauses of the charter party the principal clauses and relevance of which are included for reference.

Voyage charter (example clauses)

The shipowner agrees to provide the vessel and stipulates her position and class. The owner also agrees to dispatch the vessel to the loading port without undue delay and guarantee the ship is in a suitable condition to complete the voyage in question and effect transport of the goods to the specified destination.

The charterer provides the cargo and agrees to pay the freight based on either weight or space. The rate of demurrage is stated and time provisions for the operations of loading and discharge of cargo are stipulated.

In the event of collision, a 'both to blame' clause is often included. This clause is written in conjunction with the Brussels Collision Convention 1910 where two ships are involved in collision, their liability being in proportion to the respective degree of fault of each vessel.

NB. US legislation is against this principal and supports any party with a cargo interest to claim 100 per cent of their loss from the non-carrying vessel, in spite of the fact that each vessel is only partly to blame. However, the owners of the non-carrying ship can recover the proportional amount paid to cargo owners from the carrying vessel. The purpose of this clause is to permit the owners of the ship carrying the cargo to recover the costs of damage they would have to pay to their shippers. (The US courts have declared this clause 'null and void' and its inclusion now has to be questioned.)

Additional clauses could reflect war conditions, ice navigation, industrial action.

Some charter parties also include a 'deviation' clause.

Time charter (example clauses)

A clause which stipulates the details of the ship and specified time of the charter period. The shipowner specifies his agreement to pay crew wages, insurance, stores, etc., and undertakes to maintain the vessel in an efficient state. The charterer agrees to pay costs of fuel, dock and harbour dues and labour costs as well as an agreed amount for the cost of hire.

An 'off-hire' clause is also included where in the event of certain circumstances the hire cost will cease, e.g. where the ship breaks down and requires to seek repairs. The vessel goes off-hire from the time of the breakdown and returns back on hire from the time she departs the repair port.

A clause which incorporates the York/Antwerp Rules, and the implied terms and warranties. The Master takes his commercial instructions from the charterer who indemnifies the shipowner against any liability which may arise from the Master signing the bill of lading. Management and passage plan criteria would still be directed from the owners.

Delivery and cancellation clauses, which if the owners failed to deliver the vessel by the agreed time, the charterer would have the option to cancel the agreement. The charterer further agrees to conduct his business in a lawful manner, through safe ports.

Additional clauses usually incorporate arbitration, war conditions, ice navigation, and stipulated payments for commissions and negotiations.

Clause definition

Deviation

Under common law a vessel may deviate:

(a) For the safety of the adventure. An example of such would be in the event of fire aboard the vessel

and the ship deviates to a 'port of refuge'. Another example is where a vessel deviates because of bad weather.
(b) To save life at sea, e.g. responding to a distress situation.

NB. Unless the charter party specifically stipulates the case, a vessel is not generally allowed to deviate to save property under common law.

Cancellation clause

The cancellation date as specified in a voyage charter is that date which is stated in the charter party as the latest date by which the vessel must be ready to load her cargo. If the vessel is not an 'arrived ship' by this date the charterer has the right to cancel the agreement.

The Master would normally 'note protest' in the event that he might be late arriving at the designated loading port. Regardless of an expected late arrival, the Master should continue towards the loading port, on the basis that the charterer will not make his decision to accept or cancel until the Master of the vessel tenders his 'Notice of Readiness'.

In the case of a time charter if the vessel cannot make the delivery date/time, the charterer usually has 48 hours to declare his intention to cancel or continue with the agreement. Any decision of the charterer is generally based on commercial reasons.

Demurrage

This is a payment by the charterer to the shipowner and is incurred when the ship has not completed her loading or discharging in the laytime stipulated in the charter party. The rate of payment is expressed in the C/P, and is usually paid by the day and pro rata.

Notice of Readiness

The vessel is normally considered an 'arrived ship' once she is inside the commercial limits of the port. Most charter party's stipulate that Notice of Readiness must be tendered once the vessel is an arrived ship. In the event that loading has commenced and the Master has not tendered a Notice of Readiness then the absence of such a notice is irrelevant.

However, before the charterer is obliged to load, three conditions must be noted:

1. The vessel must be an arrived ship.
2. The vessel must be ready to load in all respects.

3. The Notice of Readiness must have been served on the charterer.

The Master is responsible for ensuring that the Notice of Readiness is tendered as soon as possible. This is not always possible where a vessel arrives at night, outside office hours. The notice may be tendered by radio but on such an occasion the Master must ensure that correct log entries are made and records of his action are retained.

NB. A Notice of Readiness cannot be tendered for some time in the future. It can only be tendered for the present time.

Once the Notice of Readiness has been accepted by the charterer, the loading time can commence (this may be altered by Customs or the C/P).

Masters should be aware that the Notice of Readiness states that the vessel is 'ready in all respects' and costs against the ship would follow in the event that the ship was not ready. For example a product tanker is found to have a tank which has not been cleaned. If the Master has already tendered Notice of Readiness and loading is suspended while the tank is cleaned, the vessel may be taken off charter for the period of cleaning.

Laydays (laytime)

A layday is considered as a 24-hour period which is stipulated by the charter party for the purpose of either loading or discharging cargo.

Reversible laydays allow both the loading and discharge operations to take place. If reversible laydays are used the consignee would need to be informed of the available time left and the bill of lading should be endorsed to specify this.

An alternative method of assessing laytime could be made with the use of a loading rate, e.g. 28 000 tons of coal to load at 4000 tons per day = 7 laydays.

Working days

These are defined as days when cargo work is carried out and count towards laydays. A 'working day' is from midnight to midnight, no matter what the working practice of the port might be. Saturday is considered a working day, unless otherwise stipulated.

NB. Sunday is usually an exclusion (Friday in an Islamic country) and would not count.

Express clauses in C/Ps specify other times which are 'not to count' as laydays.

Running days

Running days are defined as consecutive calendar days which run from midnight to midnight. Running days would include Sundays and bank holidays, unless expressly excluded by the C/P in which case they would not count towards laytime.

Weather working days

Where weather working days are applied for conducting cargo operations then in the event that cargo work is suspended, due to bad weather, any time lost would not count towards laytime.

Commercial documentation

The bill of lading

When goods are shipped under a bill of lading (B/L) in a general ship, the B/L will provide three functions:

1. It will act as a receipt for the goods received on board the vessel.
2. It will be seen as evidence of a contract of carriage.

    NB. It cannot be considered as an actual contract because it is only signed by the carrier and not by the shipper. However, once signed by the carrier the carrier is bound under its terms.

3. It acts as a document of title to the goods described therein and is proof that the person holding the signed B/L is the rightful owner of the goods and as such is the person entitled to receive them.

The B/L is usually issued in triplicate by the carrier. Alternatively a B/L can be issued under the charter party where it remains as a receipt for goods being carried under charter party terms, on the proviso that it remains in the hands of the charterer or the shipper. Once it is passed to a third party, i.e. a consignee, from the charterer, it provides conclusive evidence of the terms of carriage.

A B/L issued under the charter party is generally supplied by the charterer. The terms of the B/L should therefore reflect the same terms as those of the charter party.

B/L contents

A bill of lading will include the following details:

1. The ship's name, which may be changed by the carrier.

2. The designated loading and discharging ports.
3. The name and address of the shipper.
4. The name and address of the consignee (alternatively to the orders of the shipper).
5. The name and address of notification of delivery and clearance.
6. A description of the goods and the identification markings.
7. The number of packages or stipulated quantity/weight as supplied by the shipper.
8. The order and condition of the goods.
9. The number of B/Ls forming the set.
10. The date of shipment.
11. The place where the freight is payable (usually the loading or discharge port).
12. The place and date of issue.
13. The signature of the carrier or of the Master.

B/L types

Their are several types of bill of lading to cover several aspects and periods of the shipment of goods:

(a) B/L–Received for Shipment. This is issued to the shipper on delivery of the goods to the shipowner or his agent (usually a dockside authority) before the ship is an arrived vessel, and before the vessel is ready to receive cargo. This B/L will state 'Received for carriage …' but the loading date will be left open.

(b) Shipped B/L. This is a B/L which is issued after the goods have been loaded on board the vessel. This may be in the form of a converted B/L from the 'Received for Shipment' to a 'Shipped B/L' following application from the shipper.

(c) Direct B/L. This is issued in the case where goods are for carriage in one ship directly from one port to another port. No transshipment is involved.

(d) Through B/L. This is issued where carriage of the goods is by several modes of transport and involves more than one carrier. Carriage is in separate stages one of which is by sea. Although the sea carrier may act as agent for the carriage of the goods for the complete journey, he only accepts responsibility for the sea stage.

(e) Combined B/L. This is a similar B/L to the 'Through B/L' except the issuer accepts full responsibility for all stages of the journey, i.e. rail, sea, air, road. These are usually issued by a Combined Transport Operator and an example of the use of a combined B/L would be the door-to-door service of roll on-roll off transport.

(f) Bearer B/L. This is the B/L which permits the goods to be delivered to whoever is in possession of the document. Endorsement is not a requirement.

The letter of indemnity

The letter of indemnity has no legal standing in the eyes of the law, and a Master should never accept such a letter where there is a risk that the quantity or condition of goods is not as they should be. It may be the case that a Master is expected to sign only clean bills of lading, despite the fact that they should be endorsed. In return for not endorsing such B/Ls the Master may be offered a letter of indemnity which may suggest the Master is released of his obligations. This is not the case and the Master who accepts an indemnity letter could be prosecuted as being party to a fraud.

If compromised, the Master might be better advised to make out his own B/Ls and sign them before presenting these to the shipper. Whatever action is decided upon, the prudent Master would be well advised to keep meticulous records of all transactions.

Signing B/Ls

Prior to signing a bill of lading, Masters should ensure that the form is correct and the dates are in order. The 'Mate's Receipts' should also be surrendered beforehand.

If freight is pre-payable, check that this has been paid.

Ensure that the true description of the goods is on the document, and that the shipowner's rights are maintained if the vessel is under charter.

A Mate's Receipt

It is usual for the shipowner to have their own Mate's Receipt forms for issue in triplicate, one copy going to the person delivering the goods, another to the ship's agents in the port and the final copy being retained on board the vessel.

The Mate's Receipt is used as an informal receipt which is given to the shipper by the ship's Mate at the time of shipment. It will later be exchanged for a bill of lading, once the B/Ls have been formalized. Unlike the B/L the Mate's Receipt is not a document of title of ownership, although the shipowner is fully entitled to issue a B/L on the strength of anyone presenting a Mate's Receipt, unless he has reason to believe that such a person is not entitled to its possession.

Bills of lading are often made up from information contained in the Mate's Receipt. To this end it is considered important that the content of the Mate's Receipt is an accurate record of the condition of the goods and their description.

NB. The Mate's Receipt needs to be more detailed than information on a shipping or boat note.

Mate's Receipts need to be endorsed as appropriate to reflect the true nature of the cargo. If the consignment is three cases short, the endorsement on the receipt should reflect that three cases are in dispute and unaccounted for at the time of loading. Leaking containers, or bagged cargo where bags are torn, should all carry endorsements to the Mate's Receipts.

Waybills

The waybill is a receipt for goods received for carriage. It is issued in triplicate to the carrying vessel. One copy is signed and retained at the port of shipment of the goods, another is delivered with the goods and receipted at delivery, while the third is receipted by the receiver and kept on board the vessel.

The waybill is recognized as a receipt only, and not a document of title. It was accepted as a means to speed up the documentation process for transportation. These days it is used more as a document for 'in-house' transactions, for example sending items between company offices around the globe. Freight costs are arranged by separate contract and are usually pre-paid. The consignee would not be required to present a copy of the waybill, but may be required to prove identity if claiming the goods.

Logistic support for marine commercial operations

Shipping movements, in and out of ports, for the purpose of engaging in the commercial activity of loading/discharging cargoes require a level of external support. The Master from his shipboard position can only do so much, even in this day and age of high technology communications. He is distant from shoreside objectives and as such requires agents to act on his and the shipowner's and charterer's behalf.

Agents and agency

The ship's agents act for and on behalf of the shipowner during the period that the vessel is in port. The port agent makes all arrangements for the ship's husbandry including stores, bunkers, cargo handling, repairs and crew formalities. They also attend to berthing details and assist with external bodies such as Customs and immigration officers in conjunction with the ship's Master. The relationship between the agent and the Master is generally one of co-operation but with modern communications the Master's authority over the agent has waned and any major decisions required, other than day-to-day business, are often transmitted from the shipping company via the agent.

Master's position

If the ship is under charter, then the Master could be either the charterer's agent, or the shipowner's agent. In the case where the charterer appoints the Master, it would be reasonable to expect the Master to act as charterer's agent. It should be noted that any third party involved in a claim, who may not be aware of the Master/charterer relationship, could be excused for thinking the Master's principal is the shipowner.

The effect of this is that the Master cannot influence conditions of a contract to which he is not a party, i.e. the charter party.

Master as shipowner's agent

The usual situation, where the ship's Master acts as the shipowner's agent, is one where the Master has the Authority to act in the best interests of the shipowner.

A common example can be seen where the Master is responsible for the supervision of loading and discharging of cargo. As the shipowner's agent, he has authority over the stevedoring and as such is responsible for any negligence on the part of the stevedore.

In comparison, if the charterer nominates and employs the stevedores then the charterer would be responsible for any fault of the stevedores.

Commercial: cargo requirements

Definitions

'Free alongside' is defined as the time the shipper delivers goods, under the charter party, to within reach of the ship's loading equipment. Any further costs for loading, stowing or trimming are then borne by the shipowner.

NB. A similar situation exists at the point of discharge, where the owner covers the cost up to the point of landing the cargo on the quayside at the port of destination.

'Free discharge' is an expression within the charter party, to signify that the consignee will bear the full costs of discharge and delivery of the goods. The expression avoids disputes between discharge and delivery costs by the parties involved.

'Free in and out' is a term which is generally employed where both the loading and discharging take place at the charterer's own establishment. It is an understanding that the charterer will pay full costs for the loading, and the consignee will pay full costs for the discharge.

'Free on board (f.o.b.)' is a contract by which the seller of goods delivers the cargo free on board the ship and pays all the costs up to the time of shipment. From this moment on, the buyer of the goods takes full responsibility for all costs including, freight, insurance, import dues and any landing charges.

'Cost insurance and freight (c.i.f.)' is a contract where the seller ships goods on board a vessel of his choosing and pays the freight, together with all costs up to the point that they are loaded on board the vessel. He would insure the goods while in transit and supply the buyer with all documents to enable the goods to be imported. The buyer would be responsible for any loss or damage after the goods have been delivered to the carrier. He would also pay Customs duty and all other expenses on arrival at the port of destination.

Cargo damage

In the event of cargo damage being incurred during the shipment period, the Master should 'note protest' and inform the consignee as soon as practical. Depending on circumstances, the Master should also arrange for a hatch survey to take place and open up the hatches in the presence of a cargo surveyor nominated by the consignee and a surveyor nominated by the ship's owner.

Surveyors should pay particular interest to the nature and cause of the damage and note whether damage has been caused by the ingress of water and if so, whether that water is fresh or salt. They should note the ventilation system is functioning correctly and suitable dunnage arrangements have been made. They should be concerned that suitable stowage arrangements throughout have been observed. In the event that further damage is found on discharge, the surveyors would be recalled at that time.

NB. Under the terms of the bill of lading, the carrier may not be liable for certain categories of loss or damage to cargo and in some cases they can limit their liability. As such the cargo owner has an insurable interest in the goods and may wish to take insurance against the goods being lost or damaged.

Dangerous goods

For additional reference see the Merchant Shipping Act (Dangerous Goods and Marine Pollutants) Regulations, based on the IMO convention, on the carriage by sea of goods which may be hazardous to people or the environment.

These rules apply to UK ships and to ships loading/discharging in a UK port.

There is a duty of the shipowner/Master, as an employer, to ensure that there is no risk to the health and safety of any person who is called on to handle, carry or stow such goods. The obligations of the shipowner/employer are further extended to such that he should provide information and training to ensure the safety of all employees.

It is also the duty of each employee to take reasonable care so as not to endanger himself or any other person when involved with such dangerous goods.

Packaging, marking and labelling of dangerous goods

For additional reference for specific cargoes see the IMDG Code.

Any dangerous goods received for shipping should be clearly marked and identifiable in accord with the IMDG code. Goods should be durably marked and provided with hazardous placards or labels. The cargo should be stowed with reference to the IMDG code and segregated from other non-compatible cargoes.

Where the goods are shipped in a container or other similar vehicle the shipowner or the Master should be supplied with a signed packing certificate stating that the packing has taken place in accord with the recommendations of the IMDG code. (The container must also be marked by a hazardous placard or label.)

Documentation for dangerous goods

Prior to taking dangerous goods on board the vessel, the Master or shipowner must be supplied with a Dangerous Goods Declaration.

This declaration should provide the correct name of the product, the UN number if appropriate, and to which class of dangerous goods the product belongs. If it is a flammable bulk liquid, then the 'flashpoint' should also be included in the declaration. The specific total number or quantity of the goods should be stated and the packaging should comply with the requirements of the IMDG code.

It would be the Master's responsibility to ensure that such goods are identified and marked on the vessel's stowage, cargo plan, and that they are identified as hazardous products on the cargo manifest. Additional documentation may be supplied with the goods, e.g. precautionary notes for dealing with spillage. These would normally be kept readily available for use by the officers and crew to provide means of dealing with spillages in a safe manner.

The Master should ensure that copies of relevant documentation regarding dangerous goods are deposited ashore prior to the vessel sailing.

Introduction to general average

As with any commercial operation one of the main objectives is to generate profit. Shipping is no different in generating profits for owners and shareholders and at the same time providing employment and benefits to all participants within the industry. Where damage does occur to either ship or cargo, profit levels are directly affected.

Insurance, compensation and administration of claims all eat into profit margins and the ship's Master can expect to be involved at various levels. A declaration of general average related directly to the ship or cargo, would be substantiated by evidence, witness statements, surveyor's reports and the Master's report.

The Marine Insurance and the different types of policies would dictate the protection afforded in damage/cost claims for general average. The awareness of the Master, as the company's representative, to minimize claims, maintain the classification of the vessel and ensure the seaworthines of the vessel are paramount.

General average is the name given to the amount paid by all parties in a venture to pay towards any member who suffers loss. If the owner of cargo has not paid his/her general average contribution, then the shipowner has a lien on the cargo.

Following any incident it may become necessary to sacrifice property or incur expenditure to save the venture. In this event, one of the interested parties, usually the shipowner, may at some time in the future make a declaration of general average.

Typical examples of when general average is declared are:

1. When tugs are employed to refloat a grounded vessel.
2. The cost of discharging cargo into lighters to effect refloating.
3. Water damage to other cargo when extinguishing a fire.
4. Cost of jettisoned cargo, to save the ship, or control a fire.

In making a declaration of general average other interested parties are the owner of the cargo, the time charterer as well as the shipowner. Any declaration must be made before the delivery of any cargo, although shipowners may effect cargo delivery when parties to the venture provide suitable security to cover their contribution to an adjustment for general average.

The role of the Master: general average

Clearly the role of the Master following any incident is to record and collect any evidence directly related to the occurrence. The actions of the Master are expected to reflect good seamanship practice first and foremost, but it should not be forgotten that the report from the Master on the sequence of events will be considered crucial.

When salvage is involved any actions by the salvor need to be closely documented together with any relevant equipment employed. Salvage awards are based not only on what property is recovered but also on the level of risk involved in the recovery.

The adjustment for general average is based on survey reports which in turn are based on surveyor's findings. Witness statements, ship's records, and relevant stability calculations could all form part of the final survey report. Masters are well advised to ensure that calculations are worked through by either a company or shipboard representative and that copies of calculations/statements, etc. are retained by the Master and/or shipowner.

Conditional requirements for general average

All parties to the venture stand to gain from a declaration of general average. However, before any act of general average can be established:

(a) There must be a common adventure.
(b) That common adventure must be in peril.
(c) There must be an expenditure of money or sacrifice of property made.
(d) The sacrifice of expenditure or property must have been a reasonable intention for the sole purpose of saving the common venture.
(e) Such expenditure or sacrifice must be considered extraordinary, e.g. mechanical breakdown during normal trading would not be construed as cause for general average, but engines damaged while trying to refloat a stranded vessel could be considered a general average act.

General average sacrifices

In addition to the jettison of deck cargo or underdeck cargo, a sacrifice by the vessel for consideration of general average could be the slipping of anchors and cables or the deliberate cutting away of rigging which could endanger the vessel's progress.

General average expenditure

With any act or sacrifice of general average, costs incurred could be considerable. Examples of note are where the cargo of a stranded vessel is discharged in order to effect refloating of the vessel. Similar discharge at a port of refuge carried out to gain access to instigate repairs would also be seen as an expense for consideration. Another common expenditure is the hire of tugs to assist in refloating a stranded vessel.

General average and deck cargoes

If deck cargo is carried with all party agreement and is subsequently jettisoned for the common safety of the venture, then the loss could be considered for general average. However, only those who are party to the agreement for the carriage of deck cargo would be affected.

Where goods are carried as deck cargo and also carried at the 'shipper's own risk' then these would not be covered by general average and only the shipper of the jettisoned deck cargo would bear the loss.

Average adjuster

The 'average adjuster' is a person who has expert knowledge of the law, relating to insurance and average assessment. He/she acts as an arbitrator and decides what is and what is not allowable in general average in accordance with the law and any established contract. The assessment known as the 'average statement' could take considerable time and would place contributory values reflecting the general average allowances.

Average bond: Lloyd's average bond

The 'average statement' may take several months to prepare and this delay could present the cargo owner with a problem in as much as the release of the cargo could be delayed. The reasoning behind this is that the shipowner is entitled to exercise a possessory lien on the cargo, as it would be liable for a general average contribution.

In order to obtain release of the cargo, the cargo owner or his underwriters may provide security by way of an 'average bond' with or without an 'average deposit'. This effectively is a cash deposit into a joint account of the shipowner and the Lloyd's agent which guarantees the settlement of the general average charges. Any monies left over after settlement are returned to the cargo owner. (Lloyd's agents represent the interests of the cargo owner.)

NB. Alternative forms of guarantee are acceptable to permit cargo release and these may be in the form of a 'bank guarantee' or an 'underwriter's guarantee'.

General average deposit receipt

This is a standard Lloyd's form which is issued once the average deposit has been received. It identifies the cargo and the bills of lading and is required by the average adjuster.

Average adjustment: example

To establish the contributions payable by the parties concerned the values of the arrived ship, its arrived cargo, the freight earned and the value of the sacrificed cargo considered for general average, are totalled:

Arrived ship (net value)	$6 000 000
Arrived cargo (net value)	$4 500 000
Shipowner's freight	$1 500 000
Cargo sacrifice (GA)	$1 000 000
Total	$13 000 000

The contribution is calculated as a proportion of each party's stake in the venture, and of the amount of the general average sacrifice:

$$\text{Shipowner } \frac{6\,000\,000}{13\,000\,000} \times 1\,000\,000 = \$461\,538$$

$$\text{Cargo owner } \frac{4\,500\,000}{13\,000\,000} \times 1\,000\,000 = \$364\,154$$

$$\text{Freight } \frac{1\,500\,000}{13\,000\,000} \times 1\,000\,000 = \$115\,385$$

Contribution paid to owners of = $941 077
sacrificed cargo

Port of refuge

Once a vessel commences her voyage she will be on route to a specific destination. If, following extreme circumstances, i.e. an accident which puts the common venture in peril, the ship may be obliged to deviate to another port or return to the same port of departure.

This unexpected destination is known as a 'port of refuge' and can be considered for the purpose of general average.

The emergency circumstances must be genuine and fulfil the rights of insurance and other contracts to:

- effect necessary repairs,
- bunker, provided that the vessel left her last port with normal, adequate reserve bunkers.

Master's communications

When on route to the port of refuge, the Master would expect to inform:

(a) The ship's owners to advise of the circumstances and the selected port of refuge.
(b) The Port Authority to request free pratique.
(c) Customs Authority to enter inwards.
(d) A proper officer to 'note protest'.
(e) Underwriters or Lloyd's agents to give notification of the accident (usually communicated by owners).
(f) Owners on arrival.

Master's finance responsibilities

The Master would be expected to familiarize himself with the costs incurred when involved with a port of refuge.

1. The amount of cargo damaged or suspected of being damaged.
2. The amount of cargo, fuel, stores, etc., discharged, considered necessary to effect repairs (would also be covered by a registered cargo surveyor).
3. Surveys to the ship's hull and machinery, carried out by a Classification Society surveyor, together with tenders for repairs.
4. All port costs: light dues, berthing fees, and pilotage costs.
5. Reloading or storage of cargo, stores or fuel, together with any insurance costs.
6. Costs incurred if vessel has to proceed to a second port of refuge.
7. Towage costs to a second port, or the cost of trans-shipment of cargo.
8. Maintenance costs of Master, officers and crew together with consumables during any prolonged period of the voyage.

Certificate of Class/Interim Certificate of Class/Certificate of Seaworthiness

From the time of building under Lloyd's Register Rules, a ship will be issued with a Certificate of Class, and provided the vessel remains without damage or other structural alteration she will continue to operate under this certificate. It is not uncommon practice when a vessel requires a port of refuge that her circumstances change and necessitate repairs. When these repairs are completed to the satisfaction of the Classification Society surveyor, he will then issue an Interim Certificate of Class.

In the event that no Lloyd's surveyor is available, a private surveyor may be employed as recommended by a Lloyd's agent or by the agent of the shipowner's P & I Club. Alternatively, two British shipmasters may be appointed to carry out the survey. However, in these circumstances, a Certificate of Seaworthiness should be issued. (In the case of damage to machinery a shipmaster and a Chief Engineer would suffice.)

Insurance underwriters would accept either certificate as being representative that the vessel is in a seaworthy condition, and allow the ship to proceed on her voyage.

NB. It should be noted that in order to remain in class, a Classification Society surveyor would be called in at the earliest possible time in order to confirm seaworthiness and subsequently issue an Interim Certificate of Class.

Noting protest

It is often the case following an incident affecting the well-being of the ship or its cargo that the Master is called on to note protest. This should be made as soon after arrival in port, and in the case of 'cargo protest', before breaking bulk. The statement, made under oath, is to a notary public or a British consul and regards an occurrence beyond his control, which has, or may give rise to, a loss or damage during the voyage.

The Master would be barred from claiming the cargo's contribution in general average if he fails to note protest within 24 hours and also notifies the consignee that he has noted protest.

In the UK, protest cannot be accepted as evidence, but in many other countries, especially on the continent, the note of protest can be admitted as evidence in a legal tribunal. The outcome of a case could depend on protest being made.

Examples of noting protest

1. In all cases when general average is declared.
2. When the ship has encountered heavy weather which may result in damage to cargo.
3. When damage is caused or is suspected of being caused for any reason.
4. In the event of a serious breach of the charter party terms by the charterer or his agent, e.g. loading improper cargo, refusal to accept bills of lading in a correct form, delaying loading without good cause, refusing to load cargo, etc.
5. In the event cargo is shipped in such a state that it is likely to deteriorate during the course of the voyage (bills of lading would normally be endorsed to reflect the condition of cargo at the time of loading).
6. When the consignees fail to discharge cargo or take delivery and remit freight in accordance with the charter party or the bills of lading.

7. When cargo has been damaged by a peril of the sea, e.g. inadequate ventilation imposed by weather conditions.

The main reason for making protest in the UK is usually to support a cargo owner's claim against the underwriters.

Extended protest

There might be occasion when the Master is unsure as to the total extent of damage incurred or the full nature of that damage. In such cases, when there is a possibility of finding more damage than is apparent, the Master should extend the protest. This can be carried out by noting protest in one port and if necessary extending protest in another port. The time limit must be within 6 months of noting and extending.

Extended protest falls into two parts:

1. A stated account of all material circumstances which led up to the incident. The statement would be signed by the Master and witnessed by another officer and would be supported by any form of documentary evidence, i.e. log books, photographs, weather charts, etc.
2. The protest made by the Master and the notary against the accident and losses/damage incurred by it.

Marine insurance

Although the shipowner is under no legal obligation to insure his vessel or the cargo owner to insure his goods in transit, it has become essential because of the high capital values involved and the reputation of operators. A contract of marine insurance is termed a 'policy' where the insurer indemnifies the assured party against loss during the marine adventure. In return the assured party will pay the insurer an agreed amount of money for the risk, known as the premium.

Shipowners would normally arrange insurance through a broker who organizes the contract through a marine department of an insurance company or directly with the Corporation of Lloyd's. Most marine policies are arranged through Lloyd's underwriters and the risk is generally spread between different insurers which may include the 'Names' in a Lloyd's syndicate.

Policy types

Time policy
This type of insurance policy provides cover for a specific period of time and is usually used for hull insurance.

Voyage policy
This type of policy is usually used to provide cover for goods in transit and the limits of the policy cover the named ports of the voyage. It is sometimes used to cover freight insurance but rarely used for hull insurance.

Mixed policy
This policy incorporates the principles of both the time and voyage policies. It is generally used for insuring the hull, machinery and equipment on board for a named voyage and covers a stated period to beyond that time of arrival at the destination.

Valued policy
This is a policy which provides a fixed and agreed insured value on the subject matter being insured. The value, once agreed, is binding on both parties to the agreement. The insured value may not always be the same as the actual value but once established for the purpose of the contract, the figure is fixed and is not open to dispute at a later time.

Unvalued policy
This policy is usually employed for freight insurance. The value of the subject matter for insurance is not specified and is established in conjunction with the Marine Insurance Act.

Floating policy
This policy provides advance insurance cover for cargo intended for shipping for an undefined period of time. The policy is established by a substantial lump sum insured. As shipments are moved forward, they are declared to the underwriter who should endorse the back of the policy. The insured value of each shipment subsequently reduces the assured sum until it is exhausted. The period of cover is therefore subject to the number and value of shipments which affect the initial lump sum.

Open cover
This is probably the most popular form of insurance cover used by operators involved in the regular import and export business. It is a long-term agreement, which is binding in honour only, where the assured party declares the nature of shipments, and the underwriter accepts these shipment details within the scope of the open cover policy. The agreement is set in general terms usually for a period of 12 months and may not always include the named carrier, because at the time of initiating the contract the carrier may not be known.

Conditions of marine insurance

Any contract of marine insurance is established on the terms of 'utmost good faith'. This effectively means that the assured should provide all relevant details to the insurer, to enable a fair premium to be established. Such representations are declared during the negotiation of the contract and set into the contract as preconditions, more commonly known as 'warranties'.

Expressed warranty

An expressed warranty is a clause entered in the contract which expresses an actual condition/requirement, e.g. that the vessel would not proceed north of latitude 75°N, or south of latitude 65°S.

Implied warranty

The implied warranty is one where it is assumed that certain conditions are taken for granted. In the case of marine insurance an implied warranty is where the vessel is seaworthy to complete the intended voyage or where the vessel is engaged on a legal venture. Clearly, the insurance company would not want to be party to an 'illegal venture' and the contract would be so established, with utmost good faith.

Insurance policy clauses

Standard clauses for marine insurance are included in the majority of policies. These were drawn up by Lloyd's in 1983 and cover the aspects of hull and machinery. A selection of standard clauses are listed below.

Navigation
Such a clause stipulates the type of navigation that the vessel could be involved in and still remain covered by the policy. Navigation with or without a marine pilot would be permitted as would be trial trips. The vessel would be allowed to tow vessels in distress, but would not be allowed to be towed herself, unless she herself is in distress or as is customary for the circumstances. The contract may also specify that the ship should not engage in towage or salvage by a previously arranged agreement. Other statements could cover a final voyage prior to the ship being scrapped. In such a case the insured value of the vessel would be the established 'scrap value'.

Termination
The policy would automatically expire if the Classification Society was changed or the vessel had her class suspended or discontinued. It would also terminate in the event of a change of owners, a change of flag, or if it was transferred to a new management or the use of title was altered.

Breach of warranty
In the event of a breach of warranty, provided that the underwriters are given immediate notice to enable the policy to be changed and a new premium set, the vessel can remain on full cover.

Such examples of breach of warranty are any change of cargo or trade, or attempting salvage or towage, or a change in destination.

Continuation
In the event that the vessel operates beyond the term of the contract, continues at sea, in distress or in a port of refuge or even a port of call, she will be retained on full cover at a pro-rata monthly premium, provided previous notice has been given to the underwriters.

Pollution hazard
This clause covers the loss or damage to the vessel which has been deliberately caused by the orders of a government authority to prevent or reduce a pollution hazard.

Assignment of the policy
If the policy is to be transferred to another party, the assured must sign a dated notice of this new assignment, and append it to the policy.

75 per cent collision clause
Following a collision with another vessel, the insurer is liable to pay 75 per cent of the costs of:

- loss or damage to another vessel,
- delay or loss of use of another vessel,
- contracted salvage of another vessel or general average.

NB. The remaining 25 per cent is usually covered by the owner's P & I Club.

General average and salvage clause
The policy covers the assured party's contribution to general average, if applicable.

Where a vessel is engaged on a voyage in 'ballast' the owner has the right to claim for sacrifice or expenditure incurred under general average and for particular charges under particular average, e.g. expenditure could be considered as costs of a port of refuge, or crew wages.

Duty of the assured

It is considered the responsibility of the assured party to take all reasonable actions to avert or reduce any loss for which a claim may arise. To encourage such action by the assured, the underwriters would contribute to any particular charges incurred to protect and save the property insured.

Notice of claims

Where a loss or damage has occurred the underwriters must be notified prior to survey. If the incident occurs abroad, then this notice would be passed to the nearest Lloyd's agent. Such notification allows a surveyor to be appointed for and on behalf of the underwriters. It also provides the underwriter the time needed to obtain tenders for repairs and designate a port to carry out such repairs. Generally, it is usually the owners who arrange the tenders, but the right of the underwriters to obtain their own tenders or seek alternative tenders remains.

NB. A 15 per cent deduction from the claim can be made by the underwriters if the owners fail to comply with the stated conditions.

A marine insurance policy would carry many more specified clauses to cover a variety of exclusions, e.g. war, capture, civil disturbance, arrest or detainment, lockouts, sister ship salvage, or loss caused by a selection of weapons, although 'war risks' are usually covered by P & I Clubs.

Peril of the sea

Probably the most influential clause of the insurance policy covers loss or damage caused by named perils. These fall into the following categories:

1. Perils of the seas, and associated navigable waters, include accident or casualty from collision, stranding, or heavy weather, but do not cover normal wear and tear from wind or wave action.
2. Fire or explosion.
3. Jettison of stores or equipment overboard in order to lighten, or relieve a vessel in an attempt to refloat a stranded vessel or right a listed vessel.
4. Violent theft by persons external to the vessel.
5. Piracy, applicable also to plunder caused by mutinous passengers or rioters who might attack the ship from shoreside positions.

6. Contact from aircraft or objects falling onto the vessel from land, dock or harbour installations, e.g. dockside crane falling onto a vessel.
7. Earthquake, lightning, or volcanic eruption.
8. Breakdown or accident to nuclear installation or reactors.

The insurance policy would also cover such items as:

- Negligence of the Master, officers, crew, or marine pilots.
- Negligence of contract repairers or charterers, provided they are not the assured.
- Latent defects in the machinery such as boilers bursting or propeller shafts breaking.
- Barratry by the Master, officers or crew.

NB. Barratry is any deliberate wrongful act which is detrimental and against the owner's or charterer's interests, e.g. selling the vessel fraudulently, wilfully running the ship aground with fraudulent intent, selling ship's equipment or cargo illegally or engaging in smuggling.

Insurance of cargo

In certain circumstances, under the terms of the bill of lading, the carrier is not liable for loss or damage to cargo and as such the cargo owner has an insurable risk. In 1982 Lloyd's drew up a set of standard clauses for use with the marine cargo insurance policy. The clauses are designated into three groups A, B and C.

The A clauses cover all risks, whereas B and C clauses cover particular risks (B and C clauses are identical with the exception of specified risks covered).

The period of insurance is defined by transit of the goods, cover being provided from the time the goods leave the warehouse and is continuous until the goods are delivered to the consignee's warehouse or suitable storage place at the named destination.

Incurred losses of marine insurance

The type of loss incurred and the subsequent handling of any claim will vary, but the principles of settlement are such that the assured party is not permitted to make a profit from the loss:

- Where a 'total loss' is agreed, the assured party must abandon what is left of the item insured to the insurer.
- Once the loss has been settled, the insurer is entitled to take over all claims and rights of the assured against any third party for damages in respect of the loss incurred. This would effectively prevent the assured party claiming twice for the same loss. This is known

as 'subrogation' and is the substitution of one party for another as creditor.

Total loss and constructive total loss

A total loss occurs when the item insured is totally destroyed or irretrievably lost.

A 'presumed total loss' occurs when a ship is missing and her total loss can be presumed following a reasonable period of time. A presumed loss is usually due a peril of the sea except in a war zone area, when it would be assumed to be a war risk.

A 'constructive total loss' occurs when the item insured is abandoned because total loss is likely, or when the cost of repairs would cost more than the insured value. The assured may keep possession of the item and claim 100 per cent partial loss or alternatively give full possession to the underwriters and claim total loss. If such is the case the insured must serve a 'Notice of Abandonment' on the insurers.

Particular average

A claim for 'particular average' occurs when a partial loss of the insured item is caused by a peril which it is insured against and which is not designated as a 'general average loss'.

A Master is advised to 'note protest' when a case of particular average is declared and call in a ship's surveyor as soon as is practicable. The need for a surveyor can be easily recognized if it is understood that particular average directly affects the party who has an interest in the insured item, e.g. the ship, the shipowner or the underwriters of the shipowner.

Typical examples of where particular average would be declared are:

- Damage to the ship due to collision or stranding.
- Damage to the hull or machinery through fire or heavy weather.
- Loss of shipboard equipment due to bad weather.
- Damage to cargo through collision.

NB. Damage to cargo could still be considered as particular average and would become the responsibility of the cargo owner (or his underwriters) as the interested party.

The average adjuster would expect to see all damage reports related to the incident, in order to differentiate between what is general average and what is specified as particular average. Additional documentation would also include any surveyor's reports on the subsequent damage.

3 Shipboard management

The managerial role of the Master is one covering many aspects, from man management to particular environmental and day-to-day safe operations of shipboard activities. Clearly the introduction of the International Safety Management (ISM) code will continue to influence the operational duties carried out aboard the vessel, and all seagoing personnel can expect to be a party to the code, no matter what rank.

The International Safety Management code

The safety management code is a mandatory requirement for ships engaged on international voyages and came into effect on 1 July 1998. (For additional reference see MGN 40(M), IMO Resolution A.741 (18) and the amendments to SOLAS 1974, Chapter IX.)

The purpose of the code is to provide an international standard of safe management for the operation of ships. Its main objectives are essentially to ensure the safety of life at sea, prevent injury, and avoid damage to the environment and to property.

The code is expressed in the broadest of terms and has a widespread effect at all levels of management, including shipping companies' management policies and shipboard practice. The ISM does not replace the need to comply with existing regulations and Masters should note well that:

> *Nothing in the instructions removes from the Master his authority to take any steps and issue any orders, whether or not they are in accordance with the instructions, which he considers are necessary for the preservation of life, the safety of the ship or the prevention of pollution.*

The verification of the ISM code will be made by the Maritime and Coastguard Agency (MCA) who will conduct audits on board ships and issue appropriate certification to the company in the form of a Document of Compliance (DOC) and to each ship in the form of a Safety Management Certificate (SMC).

How it works

As with any operational policy, it is only as good as the men and women who are active within its perimeter. Many could easily dismiss the code as another paper work exercise, and proceed in ignorance of the basic need for the code. Mariners worldwide have always been conscious of the need for a strong safety attitude in everything they do. Despite this, some 70 per cent of accidents at sea can be directly linked to human error, and the remainder have a human element associated with the cause.

One of the main functions of the safety management code is to establish a Safety Management System (SMS) as a part of the basic infrastructure of both the company and the individual ship. Once in place this management system will provide:

1. A safe environmental protection policy.
2. Instructional and procedural systems to ensure the safe operation of ships.
3. A clearly defined chain of command to include shore-side personnel as well as shipboard personnel.
4. Defined levels of authority and lines of communication between all personnel.
5. A reporting procedure for accidents and non-compliance with the code.
6. Emergency response procedures for all marine emergencies.
7. A procedure for the conduct of audits and management reviews.

The above is not meant to impose further burden on seafarers. What the code is doing is forcing ships and companies to establish *in situ* procedural checklists to take account of routine operations on board the vessel and ashore to ensure that correct responses are made and essential actions are not overlooked.

A typical example of this could be seen with the checking of the bridge/engine room and essential navigational equipment prior to sailing. Many vessels now carry out this checking procedure by means of a comprehensive checklist.

It has been obvious throughout history that routine activities are all well and good, and must be seen to be working, but the code must also take into account every conceivable marine emergency. Incidents such as collision or grounding, fire or pollution have continued to occur in the past and have not always been handled as professionally as one might expect. If the infrastructure was in place before the incident and a checklist of necessary actions was available then possibly the correct responses might more easily drop into place.

NB. This should not be taken as dictating what action the Master should take, and in no way takes away the Master's authority in the event of an emergency.

Designated person(s)

In order to provide a link between the company and those persons on board the vessel, every company should designate a person or persons on shore who has direct access to the highest level of management.

Such a designated person would have the responsibility and the authority to monitor all safety and pollution prevention aspects of each of the company's ships. The designated person would also ensure that adequate resources are available, together with shore-based support as required.

Master's responsibilities

The company must define and document the Master's responsibilities regarding:

- The implementation of the safety and environmental protection policy of the company.
- Motivation of the crew in observing and carrying out the policy.
- The issue of appropriate orders and instructions in a clear and precise manner.
- Verification of specified requirements.
- Review of the Safety Management System which is in operation.

The company policy on board the vessel should include a statement emphasizing the Master's authority and the Safety Management System should stipulate that the Master has the overriding authority and responsibility for the safety of the vessel.

Company responsibilities

In implementing the code, the responsibility of the shipping company must establish on-board procedures for any marine emergency and the expected activities to provide corrective action. There must be in place an established programme for emergency drills and exercises to prepare for emergency actions, and the SMS should provide organizational procedures for dealing with any shipboard emergency.

The function of the Safety Management System (SMS)

The SMS should be the vehicle by which hazardous situations can be avoided. Generally, regular inspections would reveal potential hazards and a comprehensive format for the inspections should be within the SMS. Once the SMS has identified equipment or system failure then appropriate corrective action can be taken.

Such inspections should be incorporated into operational and planned maintenance schedules and as such be recorded with faults and corrections fully documented. The control of documentation should also be an ongoing function of the SMS in the form of a safety management manual, relevant to the vessel with obsolete documents removed and valid documentation kept readily available for audit and review.

The SMS should cover all activity aboard the ship whether at sea or in port and in particular:

1. General shipboard operations, e.g. standing orders, mooring operations, fire and security, maintenance, safety committee, etc.
2. Shipboard operations in port, e.g. embarkation of passengers, loading and securing cargo, bunkering, gas-freeing, etc.
3. Preparation for sea, e.g. stability assessments, testing engines and navigation equipment and watertight doors, verification of personnel on board, etc.
4. Shipboard operations at sea, e.g. watchkeeping, navigation, communications, equipment monitoring, etc.
5. Emergency and contingencies, e.g. fire prevention, damage control, equipment failure, passenger control, pollution control, etc.

See Figures 3.1 and 3.2.

Company obligations under ISM

For the safety management code to be effective, it is essential that the shipping company provides practical support to the ships and the men and women who sail them, especially to the Master of the vessel.

It is expected that the company should ensure the manning of its ships are in accordance with the regulations and in particular:

Figure 3.1 Example of shipboard operations: planned maintenance ongoing, aloft the deck of a tanker vessel while at sea in calm conditions

(a) The Master is properly qualified and in possession of a valid certificate.*

(b) The officers and crew are certificated and medically fit to carry out seagoing duties.†

(c) That new personnel or persons transferred to other safety duties are familiar with their new surroundings, prior to sailing.

(d) That the Safety Management System (SMS) of the vessel is understood by all persons on board, together with relevant instructions and regulations in effect.

(e) That recognition of relevant training needs are identified, and that personnel are provided with a training programme to meet the necessary requirements.

(f) That updated information and changes to procedures are received by the relevant personnel, in a language that they can understand.

(g) That ship's personnel can communicate effectively in the execution of their duties.

It must be considered an essential element of the code that the company ensure that the Master of the vessel is fully conversant with the SMS operated aboard the vessel and that the company would support the Master in the safe performance of his/her duties.

Personnel management and ISM

The introduction of the ISM and the operation of Port State Control encompasses all areas of shipboard operation and inspections rely heavily on the quality of not just equipment resources but also the safety culture of the crew involved. To this end, greater attention must be paid to man management procedures, employment law and the Health & Safety at Work Act.

There is a need for principal officers to follow approved policies not only to satisfy the requirements of current legislation but because there is a moral need to achieve mutual respect inside the working environment. Management styles ashore and afloat differ considerably,

---

* Qualified personnel as per the Standards of Training, Certification and Watchkeeping (STCW '95).

† Personnel should also be familiar with the type of vessel and the trade in which it is engaged and be in possession of any relevant associated qualifications in order to carry out their duties effectively.

Figure 3.2 Example of typical watchkeeping duties while the vessel is at sea: a Watch Officer monitors the fire detection and alarm system. (Fire detection system manufactured by AFA Minerva Ltd Marine Offshore Division)

but both require quality leadership and a degree of discipline.

Control of the shipboard environment, because of the associated hazards with the position, requires a positive approach to essential discipline. Many companies operate a code of conduct to which a recognized degree of fairness is attached. However, senior officers should be aware that all the regulations in the world are only credible if all parties abide by them and the rights of the individual are retained. The alternatives could lead to litigation and compensatory claims being made against companies or individuals.

Crew (employment) agreements

For additional reference see SI 1991 No: 2144, Crew Agreements.

It is the direct responsibility of the ship's Master to ensure that the agreement of employment (or copy of the agreement), once approved by the MCA, is displayed in a prominent position for access by every seaman who is a signatory to it. Once the agreement is ended, i.e. at the completion of a voyage or when seamen have been paid off, the agreement is returned to the proper officer within three days, at the port in which the voyage ends.

NB. A proper officer being a shipping officer or marine superintendent at the port.

The agreement is usually one of two standard forms, either:

(a) ALC (BSF) 1(d) – a British federation form.
(b) ALC (NFD) 1(d) – a non-federation form.

These may be adopted by any ship and amended, subject to approval of the MCA.

Crew lists

When opening up a crew agreement, it is a requirement that any ship, other than a fishing vessel or yacht, draws up a crew list. This list must contain all persons on board the vessel who are party to the agreement including the Master, any juveniles together with any person who may be exempt from signing the agreement.

The crew list should contain general particulars of the vessel and the name and address of the ship's owners. Each person being listed should also have detailed information recorded: name, address, rank or grade, discharge book or passport number, detail of last ship, date of joining and the date and place of discharge, and the reason for discharge. The details of individual's next of kin would also be noted.

The shipowners are required to keep a copy of the crew list at the normal UK address from which they do business. In the event of a marine accident where the ship is lost they would be expected to deliver the list to a marine superintendent.

The Master would therefore be expected to keep the crew list updated and inform the owners of any changes within three days of such a change.

Safe manning requirements

Since 1992 it has been a requirement that all UK ships and other seagoing vessels in UK waters have a Safe Manning Document (see SI (1992) No. 1564).

The subject of the safe manning of ships at sea is a direct concern to the level of seaworthiness of the vessel and its intended trade. The vessel is subject to detention by the Marine Authority if its inspectors have reason to believe that a vessel is undermanned.

The question of what constitutes 'safe manning' is then considered and each vessel must have sufficient numbers of appropriately qualified officers and crew to conduct seagoing operations in a prescribed safe manner. Officers and crew must:

1. Be able to maintain safe bridge watches at sea in accordance with the requirements of the STCW and ISM regulations (under any normal or irregular condition).
2. Conduct mooring and unmooring operations for the vessel in an adequate, safe manner.
3. Maintain and operate all fire-fighting and life-saving appliances in a designated manner.
4. Maintain and operate all watertight closures in order to secure the ship's hull.
5. Muster and control the disembarkation of passengers and non-essential personnel aboard the vessel.
6. Muster damage control parties to tend emergency situations.
7. Manage all safety functions when the vessel is at sea.
8. Conduct and maintain a safe engineering watch or machinery surveillance operation while the vessel is at sea.
9. Maintain and operate main and auxiliary machinery to be able to overcome expected perils of the sea.
10. Maintain cleanliness and safety levels in machinery spaces so as to avoid fire risk.
11. Provide a level of medical care on board the vessel.
12. Operate and maintain a safe radio communication watch.

Approval for safe manning of the vessel also requires that the vessel is adequately equipped with appropriate life saving appliances to satisfy the regulations.

Masters should also note that concession on sailing short handed with either a Deck or Engineer Officer short is permitted, provided that watchkeeping activity is not adversely affected. A vessel may also be permitted to sail if a rating is short, provided that a full complement of officers is on board. (Periods where an officer is missing from the complement cannot be allowed to run consecutively.)

In the event that the Master or second-in-command is short, such a concession would only be permitted to the next port of call.

When considering 'safe manning' the authority is empowered by the Secretary of State to recognize certificates and qualifications issued by another state as being equivalent to UK qualifications.

Verification of safe manning levels

An authorized person may inspect the vessel to verify the safe manning level on board and may also check the skills and ability of those persons on board to assess their capability to function in a correct capacity.

Shipboard safety

The safety organization aboard the vessel must encompass all persons on board and have a system that allows two-way communication between various management levels and active members of the crew (see the Code of Safe Working Practice, Chapter 4).

Although the employer is ultimately responsible for the safety of all persons on board the vessel, the immediate responsibility lies with the ship's Master as the company's representative. Under Merchant Shipping Regulations applicable to a UK registered ship, where a crew of 5 or more are employed, the vessel must carry prescribed 'safety official(s)'.

Safety officials include: Safety Officers, safety representatives, and safety committee members.

The Safety Officer

The Safety Officer of the vessel is appointed by the employer (who may or may not be the shipowner). The position of Safety Officer cannot be resigned, and should preferably be a person who has attended a Safety Officer's/Safety Official's training course.

The choice of an appropriate person to act as the ship's Safety Officer should be made with care. Ideally, it should not be the Master or a person who is appointed to administer medical assistance on board. The reason for this is that the Safety Officer will be expected to make representation to the Master, and in the event of an accident a Medical Officer would be better employed administering medical attention rather than investigating the accident details.

When appointing a Safety Officer, the employer/Master should be mindful that the appointee must have 2 years' seatime over the age of 18.

Duties of the Safety Officer

The Safety Officer:

1. Should ensure that the provisions of the Code of Safe Working Practice and the employer's Occupational Health & Safety policies are complied with and standards of safety consciousness among crew members are encouraged.
2. Will investigate all accidents and dangerous occurrences or potential hazards affecting the vessel and her crew.
3. Will also investigate any complaint regarding the Occupational Health & Safety policy.
4. Will inspect the vessel every 3 months under the Occupational Health & Safety policy.

5. Make representations and recommendations to the Master or employer on legislation requirements as per MGNs, MINs, MSNs, Statutory Instruments, and the CSWP.
6. Maintain a detailed record book regarding: accidents, dangerous occurrences, complaints, statements and investigations, representations and recommendations together with action taken.*
7. Has the authority to stop any work on the vessel which may give rise to a serious accident. If the Safety Officer does cause work to stop the Master must be informed of the action, and it would fall to the Master to order the resumption of that work.

It is normal practice for the Safety Officer to make a report to the safety committee (if one is in operation) and generally assist the Master in the completion of any accident reporting documentation.

NB. Practising seafarers will be aware that the format of promulgating information within the marine industry changed in 1997 in as much as the MCA now issue three types of notices:

- Merchant Shipping Notices (MSNs) (previously 'M' Notices) – These convey mandatory information which must now be complied with under the law of the United Kingdom. They amplify the Statutory Instruments and as such the information they contain must be observed and considered a legal requirement.
- Marine Guidance Notes (MGNs) – A new type of notice which relates directly to specific topics affecting changes and new requirements. Example topics: MARPOL, SOLAS, ISM.
- Marine Information Notes (MINs) – These are information notices which are aimed at the service and related industries such as training establishments, and manufacturing companies supplying marine equipment. These MINs will be published with a self-cancellation date.

All the above are published by the United Kingdom Maritime and Coastguard Agency and carry one of the following prefixes:

(M) effective for merchant vessels,
(F) effective for fishing vessels,
(M & F) effective for both merchant and fishing vessels.

---

* The record book should be available for inspection by any of the following: safety representatives, safety committee, the Master, or a Maritime and Coastguard Agency representative.

The safety representative

A safety representative is an elected position, and like the Safety Officer, the person so elected must have at least 2 years' seagoing experience after the age of 18 years. (In the case of a tanker vessel, a safety representative would require at least 6 months of this sea time spent on a similar vessel.)

Safety representatives on board the vessel constitute a valid reason for the formation of a safety committee, if one isn't already established. As crew representatives they are expected to attend all the safety committee meetings.

The number of safety representatives on board is dictated by the complement of the crew:

6 – 15 crew	One elected by officers and ratings together.
16 + crew	One elected by officers and one elected by ratings.
Over 30 ratings	One elected by officers and three elected by ratings. (One each from the deck, engine room and catering departments.)

The Master must organize the election of safety representatives within 3 days of being requested to do so by any two persons who are entitled to vote in the election. He should also enter the name of every safety representative so elected into the Official Log Book.

Powers of the safety representative

1. The safety representative should participate in investigations and inspections carried out by the Safety Officer, with his agreement. The representative may carry out his own investigations after notifying the Master.
2. He represents the crew on matters of occupational health and safety at work.
3. He can make representations to the employer through the Master.
4. He may request an investigation by the Safety Officer through the safety committee.
5. He has the right to inspect the Safety Officer's record book.

Being an elected position the safety representative can resign his position if he or she no longer wishes to continue with the duty.

The safety committee

It is a statutory requirement that if the complement of the crew is of such a number as to generate a safety representative, then the ship must operate with a safety committee. The composition of the committee should be established with the Master in the role of chairman, the Safety Officer of the ship, and all safety representatives, together with any other person relevant to the business in hand. A secretary, who is preferably a non-safety official, would also be expected to record and document the committee business.

The success of the safety committee will depend on the willingness of the members and an enthusiastic attitude to make the ship's environment a safe working area. The forum is probably the most important element of the occupational health and safety organization on board because it is when all safety officials meet together in a common cause. With the Master as chairman, the committee has the power and authority to make effective decisions affecting the well-being of all persons on board.

The committee should not be so large that it becomes unwieldy; there is a danger of this on large passenger vessels with large crews. The use of subcommittees to take on separate issues could be a positive method of dealing with distinctive problems. Any matters arising from the meetings should be administered in such a manner as to bring relevant items to the attention of the crew, e.g. extracts of the minutes being posted on relevant notice boards.

Duties of the employer

It is the duty of the employer to provide access to information to safety officials, to allow them to conduct their duties in a proper manner. The safety committee and the officials should be kept informed of any dangerous cargoes on board or any hazards affecting the ship and permit inspections of the vessel under the occupational health and safety policy.

The employer must also provide accommodation and necessary equipment to allow the committee and officials to function in a correct manner. To this end the Safety Officer and safety representatives must be allowed time from routine duties to attend committee meetings and carry out their respective functions.

Shipboard authority

Command of a ship

This is defined as the direct authority to control a ship, while competence of command is expected to include the exercise of skills, knowledge and experience to command the vessel.

Ship's Master

A ship's Master is a qualified Master mariner, with the appropriate Certificate of Competency, who is charged by the shipowner to ensure the safety of the ship, cargo, passengers and crew. The position he/she holds would call on the skills of navigation and seamanship, as well as the ability to conduct business transactions, with due concern for those parties which fall within the command authority of the vessel.

Legal responsibility

This requires persons in command of a vessel at sea to exercise control of the seagoing craft to ensure that its operations fall within the law. However, as in any walk of life there is a personal responsibility which is both moral and ethical to conduct the function of command in a proper manner.

NB. Under IMO Resolution 433 (XI) the shipmaster is not constrained by the shipowner, charterer or any other person from taking any decision, which in his professional judgement as a shipmaster is necessary for the safety of the vessel or for the protection of the marine environment.

It should also be noted that the ship's Master is protected by appropriate provisions, which include the right of appeal contained in national legislation, collective agreements, or contracts of employment from unjustifiable dismissal or other similar action by a shipowner, for exercising his/her professional judgement.

Leadership skills

The ship's Master is expected to demonstrate exemplary personal conduct in the command of his vessel and provide firm and confident guidance for the ship's company to work in a congenial and socially acceptable environment.

Management and control

This is defined as a process of using personnel and resources to achieve required objectives. Shipboard activity must be monitored to ensure compliance within the law and that the company's policies are conducted in accord with correct procedure.

Vicarious liability

Vicarious liability is attached to the nature of the position held by the ship's Master, e.g. liability for marine pollution is attached to the position no matter who is at fault as to the cause of the pollution.

Accountability

This is the commercial or military process which exercises control over any actions or decisions made by any party, inclusive of a ship's Master.

Port State Control

It has long been realized that the condition of the world's merchant fleets were without the element of control which monitors deteriorating safety standards in ships. This obvious threat by substandard ships to the marine environment and to the lives of seafarers needed to be addressed. This has now been undertaken with the adoption of the Paris Memorandum in January 1982, superseding the Hague Memorandum of 1978.

Although the Hague agreement provided some measure of enforcement it did not go far enough to provide total effective control. What followed was the Paris Memorandum of Understanding (MOU) on Port State Control which provided a more comprehensive framework to encompass not only the working and living conditions at sea, but definitive implications of maritime safety and the prevention of pollution.

The conference was attended by the Maritime Authorities of 14 European nations who agreed to enforce the following instruments:

- SOLAS 1974, and the Protocol of 1978.
- Convention on Loadlines 1966.
- Convention for Prevention of Pollution from Ships 1973 as modified by the Protocol of 1978.
- Convention on Standards of Training and Certification of Watchkeepers 1978 (STCW), modified 1995.
- Convention for the Prevention of Collision at Sea 1972.
- Merchant Shipping (Minimum Standards) Convention 1976.

The practical outcomes of the agreement are such that MOU members conduct Port State inspections under the direction of a secretariat based in the Netherlands. The IMO act as observers to the organization's control committee and the results of all inspections are recorded in a database accessible to all concerned parties.

The operation of Port State Control

It is the responsibility of the flag state to ensure that its vessels operate safely and in a pollution-free manner under international regulations. However, it is recognized that international rules are not uniformly applied and it is a mistaken assumption that by notifying a flag state of a vessel's deficiencies that these will be automatically corrected. Examples of this were clearly noted when certain states quickly granted temporary dispensations in order to keep vessels operational.

With the introduction of Port State Control activity, compliance to the standards laid down in the conventions was able to be enforced by regional inspections of vessels entering into a participating country's ports. Such inspections may not always reveal deficiencies because of the time available for inspection or the state of loading of the vessel. However, what will evolve in the future, as the effectiveness of the monitoring starts to increase, will be the recognition of substandard ships and the visibility of ineffectual ship management practices.

Each participating Maritime Authority would supply 'inspectors' to ensure the effectiveness of Port State Control, and that foreign merchant ships entering that country's ports are liable to inspection and detention if found to be substandard. Different levels of defects will of course be revealed over periods of time and should not always lead to a vessel being detained. Inspectors should note problem areas and shipowners and managers should be given time to take such action as necessary to rectify problems.

In the event that defects are not resolved or the type of defects are considered too numerous and threaten the overall safety, then this could ultimately result in the vessel being detained.

Lord Donaldson's report to the UK Government regarding 'Safer Ships – Cleaner Seas' reported that 2000 foreign flag ship inspections were carried out in UK ports during 1992. Sixty per cent of these revealed a variety of deficiencies. While 6 per cent were so serious as to warrant detention of vessels in port.

During the same year 247 UK registered ships were inspected outside the UK. One in four revealed deficiencies. Six vessels had to be detained in port while faults were remedied. Of 237 Turkish registered ships that were inspected nearly two-thirds had defects and 15 were detained. Of 375 vessels registered with St Vincent and Grenadines, more than half revealed deficiencies, and one-sixth of the ships had to be detained.

Port State Control: inspections

In meeting the requirements of the MOU an appointed inspector of the country's Maritime Authority should conduct an inspection of the vessel following its arrival at the port. The first priority of the inspector is to check the validity of all the ship's documents and certificates. Should these be found to be invalid, or there are 'clear grounds' for believing that the ship's condition or that of its equipment or crew do not comply with the requirements of relevant instruments, then a more detailed inspection could be conducted.

NB. 'Clear grounds' is defined as evidence of operational shortcomings in compliance of the various regulations and instruments as stipulated by the specific conventions.

An inspector should also be expected to ascertain that all cargo and other marine operations are being conducted in a safe manner which conforms to IMO guidelines.

Following observation of an abandon ship and fire drill aboard the vessel, the inspector should check that crew are familiar with such emergency duties.

He should also assure himself that key personnel are able to communicate effectively with other persons on board, and check if the vessel has a history of failure of compliance and whether such failings of operational requirements have led to the vessel's involvement in incidents.

Carrying out an inspection should not unduly cause the vessel to be delayed and any tests of a physical nature should not jeopardize the safety of the ship, its crew or its cargo. Full inspections may, because of limited time, not always be completed in one port and could be ongoing with other member states' inspectors.

During an inspection items of particular interest should include: muster lists, damage control plans, communications, bridge operations, fire control plans, fire and abandon ship drills, movement of oil and oily mixtures inside machinery spaces, tank cleaning operations, handling of garbage, and the carriage of dangerous goods.

Defects should be noted and confirmed in writing to the Master of the vessel. The inspector should forward such details to the flag state and the information should be included in the computer database at St Malo, France.

Inspection report data

A list of inspection report data for detained vessels is given below.

Ship's Name
IMO Number
Classification Society
Flag
Deadweight
Year of Building

Cargo Type
Owner
Manager
Charterer
Charter Type
Port and Date of Inspection
Last PSC Inspection (Date and Port)
Last Special Survey
Serious Deficiencies
Action.

For an example of a form for the report of inspection in accordance with the Paris Memorandum of Understanding on Port State Control, see Figure 3.3.

Discipline on board

The shipboard environment in the Mercantile Marine has in the past been referred to as the fourth arm of the military and this became a visible fact during the Falklands War in 1982, with many merchant ships involved with logistic support for the Royal Navy. Discipline in the military is a recognized approach towards achieving objectives and a method of providing legitimacy to reaching a goal; while in the commercial arena of merchant ships, discipline is still a requirement but generally achieved with a willingness to provide a state of orderliness on board the vessel.

Society, in the main, creates this state of orderliness by obeying the law. In the confines of a ship the level of regulation and adherence to the Merchant Shipping Act generates a similar situation. Clear evidence exists that not everybody obeys the law, or there would be no need for prisons. Operating a merchant ship involves no less commitment from the crew and they are no different to other sectors of society. Consequently regulations are imposed or discipline procedures are put in place for the good of all.

In the Merchant Navy, a code of conduct was established that was applicable to all British ships and their crews. Now, company policies follow a similar procedure in order to protect themselves from charges of unfair dismissal or being unfair in their administrations.

The law is for everybody, but it is unfortunate that elements of our society seek to abuse it for their own ends. Masters must be mindful that any discipline procedure that they take must be legal and without prejudice, and it should not encroach on a person's civil liberties.

It is essential that any incident which involves a breakdown in discipline:

(a) is dealt with promptly,
(b) judgement is firm, fair and must be consistent,
(c) judgement is based on facts,

FORM A

REPORT OF INSPECTION IN ACCORDANCE WITH THE PARIS MEMORANDUM OF UNDERSTANDING ON PORT STATE CONTROL *)

(issuing authority)
(address)
(telephone)
(telefax)
(telegram)
(telex)

Copy head office
(surveyors copy)
(master's copy)
(IMO copy)

1 name of issuing authority ....

2 name of ship .... 3 flag of ship ....

4 type of ship ....

5 call sign .... 6 IMO number ....

7 gross tonnage .... 8 year of build ....

9 date of inspection .... 10 place of inspection ....

11 relevant Certificate(s)

a title	b issuing authority	c dates of issue and expiry
1		
2		
3		
4		
5		
6		
7		
8		
9		
10		
11		
12		

d the information below concerning the last intermediate survey shall be provided if the next survey is due or overdue

date	surveying authority	place
1		
2		
3		
4		
5		
6		
7		
8		
9		
10		
11		
12		

12 deficiencies ☐ no ☐ yes (see attached FORM B) ☐ SOLAS ☐ MARPOL
13 ship detained ☐ no ☐ yes
14 supporting documentation ☐ no ☐ yes (see annex)

district office .... name ....
duly authorized surveyor of (issuing authority)

telephone ....

telefax/telex/telegram .... signature ....

*) Maritime Authorities of Belgium, Canada, Denmark, Finland, France, Germany, Greece, Ireland, Italy, the Netherlands, Norway, Poland, Portugal, Spain, Sweden and the United Kingdom of Great Britain and Northern Ireland have concluded a Memorandum of Understanding harmonizing the procedures on Port State Control. This Port State Control is based upon the international conventions on safety, the protection of the environment and living and working conditions on board ships as adopted by the International Maritime Organization and the International Labour Organisation. If this inspection report does not contain any remarks under the heading 'nature of deficiency' the above Maritime Authorities will seek to avoid inspecting the ship again for a period of six months after the date this report was issued, unless there are clear grounds for another inspection. This inspection report has been issued solely for the purpose of informing the master and other port States that an inspection by the port State, mentioned in the heading, has taken place. This inspection report cannot be construed as a seaworthiness certificate in excess of the certificates the ship is required to carry.

page 17

Annex 3 — Paris Memorandum of Understanding on Port State Control

FORM B

REPORT OF INSPECTION IN ACCORDANCE WITH THE PARIS MEMORANDUM OF UNDERSTANDING ON PORT STATE CONTROL

(issuing authority)
(address)
(telephone)
(telefax)
(telegram)
(telex)

Copy head office
(surveyors copy)
(master's copy)
(IMO copy)

1 name of issuing authority ....

2 name of ship .... 5 call sign ....

9 date of inspection .... 10 place of inspection ....

15 nature of deficiency	Convention[1]) references	16 action taken[2])

name ....
duly authorized surveyor of (issuing authority)

signature ....

[1]) To be completed in the event of a detention
[2]) Codes for actions taken include i.a.. ship detained/released, flag State informed, classification society informed, next port informed (for codes see reverse side of copy).

page 18

Figure 3.3

(reverse side of Form B)

codes for actions taken:

code	
00	no action taken
10	deficiency rectified
12	all deficiencies rectified
15	rectify deficiency at next port
16	rectify deficiency within 14 days
17	master instructed to rectify deficiency before departure
20	grounds for delay
25	ship allowed to sail after delay
30	grounds for detention
35	ship allowed to sail after detention
36	ship allowed to sail after follow-up detention
40	next port informed
45	next port informed to re-detain
50	flag state/consul informed
55	flag state consulted
60	region state informed
70	classification society informed
80	temporary substitution of equipment
85	investigation of contravention of discharge provisions (MARPOL)
95	letter of warning issued
96	letter of warning withdrawn
99	other (specify in clear text)

REPORT OF DEFICIENCIES
NOT FULLY RECTIFIED OR ONLY PROVISIONALLY REPAIRED

in accordance with Annex 2 to the Paris Memorandum of Understanding on Port State Control

(Copy to maritime authority of next port of call, flag Administration or other certifying authority, as appropriate, as required by 3.8 of the Memorandum and to the Port State Control Secretariat) (see Chapter 2 of the Manual for Surveyors for maritime authority addresses)

1. From (country): .... 2. Port: ....

3. To (country): .... 4. Port: ....

5. Name of ship: .... 6. Date departed: ....

7. Estimated place and time of arrival: ....

8. IMO number: .... 9. Flag of ship: ....

10. Type of ship: .... 11. Call sign: ....

12. Gross tonnage: .... 13. Year of build: ....

14. Issuing authority of relevant certificate(s): ....

15. Nature of deficiencies to be rectified: .... 16. Suggested action:
(including action at next port of call)

....

....

....

....

....

....

17. Action taken:

....

....

Reporting Authority: .... Office: ....

Name: .... Facsimile: ....
duly authorized port State control officer of (reporting authority)

Signature: .... Date: ....

page 19

page 20

Figure 3.3 *(continued)*

(d) that any action thought necessary follows a correct legal procedure.

Masters when dealing with a breach in discipline should endeavour to stay calm and consider each case and individual separately on their respective merits. Incidents should be fully documented, and when appropriate, statements from witnesses should be collected as evidence for possible future use. When the need arises, full and complete entries would be required in the Official Log Book together with any fines imposed.

Discipline styles

Discipline in any form is a management tool to enforce organizational standards and can be either a preventive or corrective discipline style.

Preventive discipline

Any action taken to reduce infractions and encourage employees to maintain standards and follow an expected format is known as preventive discipline. The objective being to develop employees to operate under a code of self-discipline as opposed to a management imposed discipline level.

This can usually develop as a source of pride to any organization. If employees know and understand what standards are expected, they will be in a position to give their support to the system if they see it working. A Master should encourage a positive system of preventive discipline, as opposed to one which will chastise a seaman after the fact e.g. far better to put up 'no smoking' signs to prevent a fire, than to reprimand a seaman for smoking and starting a fire.

Crew members are more likely to support standards that they have helped to create and put in place, than be ignorant of what is expected of them.

Corrective discipline

This style of discipline is instrumental after the fact, once an infraction of the regulations has occurred. Its objective is to discourage further breaches in discipline with the hope that because of the punishment, compliance to regulations will take place in the future. The punishment, which is generally known as disciplinary action, is meant to reform the offending person, deter other persons from committing a similar action, and maintain effective standards for the good of all.

The objective of corrective discipline is not to punish for the past, but rather to educate and reform the culprit for the future, e.g. a seaman is absent without leave and fined two days' pay – one day for the loss of work, the second day as a punishment – with the hope that he will realize that his absence has meant his duties have to be covered by other crew members and that he will be working at least one day for nothing.

Progressive discipline

Nothing is ever black or white when dealing with a disciplinary action, and for this reason most shipping companies operate a progressive disciplinary procedure.

The objective of this system is that it provides scope for the offender to learn from his/her initial mistake and take personal corrective action to ensure that reform is achieved. It also allows time for management to correct the situation, if it is correctable.

Progressive discipline may lead to dismissal, but dismissal can be a reflection on management's failure to manage. However, this is not always the case, as some employees cannot be reformed no matter how hard people try.

An example of progressive discipline is well practised at sea in the following manner: verbal warning by supervisor, first written warning on record, second written warning, notice of dismissal, appeal by offender, dismissal.

Over recent years considerable legal implications have been imposed on employer's rights to dismiss an employee. Bearing this in mind it would be normal practice to see company involvement if the situation developed to a second written warning.

Whatever system is followed attention must be given to the rights of the individual so that employment law is not infringed.

Due process must include:

- The rights of the individual to be heard and possibly be represented by another person.
- The disciplinary action is reasonable, taking into account the nature of the offence.
- The presumption of innocence is assumed without reasonable proof.

The system is one which can and must be flexible. Clearly certain offences are so severe that they could not be handled with a verbal warning. In such a case dismissal becomes the first and only option or alternatively the law of the land would take over.

Dismissal

A breach of discipline can be of a minor nature which might warrant an informal warning or a written warning, but equally there are more serious breaches in policy that require a more immediate and positive response, such as immediate dismissal.

Such infringements could include actions like:

(a) Assault.
(b) Possession of offensive weapons.
(c) Wilful damage to the ship or property on board.
(d) Disobedience of orders relating to the safety of the vessel.
(e) Conduct which endangers the ship or persons on board.

A company policy should without doubt include many other examples but the Master must be reasonably satisfied that the offence has been committed, and the case is proved against the individual. Then dismissal could take place immediately or at the end of the voyage.

Any action taken by the Master or the company does not preclude any appropriate action which might be applicable under the rule of law.

Disciplinary action by Master: summary

The general advice to Masters who find themselves having to take disciplinary action is to take action promptly and make themselves familiar with the disciplinary procedure beforehand. Seek advice if in any doubt before taking the action and investigate the circumstances thoroughly, obtaining statements and witness reports. Advise the employee that they will be subject to a disciplinary hearing and remind them of their rights of representation.

Provide the individual with an opportunity to state their case (there are always two sides to a story), making sure that there is a disciplinary case to answer. Once action is decided, this action must be confirmed with the employee who will be reminded of their rights of appeal against your decision. Whatever the outcome, it is essential that full and complete records are kept of all stages of the procedure.

If a final written warning is given, then this should define what action will be taken in the event of any repetition of the offence.

The prevention of drug and alcohol abuse

Special note

The use of banned substances seems to be the scourge of the twentieth century. So much so that misuse of drugs is more noticeable in virtually every walk of life, inclusive of shipping. Combined with the hazardous nature of the marine environment it is wonder that concerns have been raised with regard to crew performance and exposure to the drug culture.

Performance by individuals can expect to suffer where drugs are being misused and subsequently the efficiency of the vessel is brought into question. Screening of new entrants into the industry and random testing goes someway towards monitoring standards amongst personnel, but not all marine administrations are policing in this manner and neither are checks anything like 100 per cent effective.

The revised STCW convention recommends guidance with regard to the prevention of the abuse of either drugs or alcohol. General advice suggests that prescribed maximum levels of 0.08 per cent blood alcohol level for seafarers during watchkeeping duties should not be exceeded. It is also recommended that the consumption of alcohol is prohibited up to 4 hours prior to serving as a watchkeeper.

Additionally, many governments are continuing to establish screening of ship's personnel as a matter of routine to identify abuse of alcohol or drugs by individuals. Masters should note that following an accident or incident the testing of crew members directly involved, of whatever rank, is to be expected by the Marine Authorities.

It is also pointed out that the responsibilities of the flag state are to ensure that standards of competence are maintained. To this end random testing of seafarers by Port State Control officers has become an established method of monitoring personnel and the level of quality maintained competence. Where flag states have issued their own certificates or endorsements to certificates of foreign seafarers, such endorsements can be withdrawn in the event that a seafarer is found to be incompetent. In such cases the flag state must inform the administration that issued the original certificate.

NB. Many shipping companies now exercise positive checks on new entrants into the industry which include medical tests prior to being assigned to shipboard duties.

4 Navigation and communication practice

Introduction

With the many rapid changes taking place in the areas of navigational practice the role of the Master has never been in greater demand. The basic principles regarding the keeping of a navigational watch as specified by the STCW convention, are such that the Master of every ship is bound to ensure that watchkeeping arrangements are adequate for maintaining a safe navigational watch.

Bearing this in mind and with the development of electronic charts and integrated bridge systems (see Figure 4.1) moving into all classes of vessels it is clear that the skills of the Master and appointed Watch Officers need to keep pace with changing circumstances. The onus on the Master is to ensure that bridge and engine room watchkeeping arrangements are adequate when the ship is at sea, at anchor or in port, and that Deck Officers are given guidance in carrying out their duties.

NB. This is especially needed when new or inexperienced officers join the vessel.

Navigation

Passage planning

There are now established legal requirements for planning the passage in advance of the voyage and monitoring the progress of the vessel in accordance with the 'passage plan'. The movement of the vessel from 'berth to berth', including pilotage waters, must be documented and all aspects covered with the formulated plan.

It is normal practice for the Master to delegate the passage planning duty to the ships Navigation Officer, but the ultimate responsibility remains with the Master to change or amend the plan and ensure its correctness. It must take into account the latest navigation warnings which could affect the vessel's progress and must be flexible to take account of changing circumstances, especially adverse weather.

Duties

The Officer of the Watch (OOW) has the responsibility for the safe navigation of the vessel during his/her period of duty. The navigation watch is maintained on the bridge, and the OOW should under no circumstances leave this position until he/she is relieved. The OOW continues with this responsibility, despite the presence of the Master on the bridge and Masters are advised that if they wish to take over the 'con' of the vessel and accept the navigation responsibility, this fact should be made abundantly clear to the OOW.

Equipment

It is expected practice that the Officer of the Watch would make full use of any and all navigational equipment on the ship's bridge to ensure the safe passage of the vessel. This must include engine controls and the adjustment of the vessel's speed as and when circumstances dictate the necessity to slow down, increase speed or stop the vessel.

The use of position fixing instruments should be gainfully employed throughout the watch period but Masters should ensure their officers are aware that they are an aid to navigation and will have a degree of instrument error, coupled with the human error in its application and a delay error. Both primary and secondary position fixing methods should be employed and a Master's standing orders should reflect this requirement.

Communications and GMDSS

Since February 1999 the Global Maritime Distress and Safety System (GMDSS) became effective for commercial shipping (see Figure 4.2). In order for

Figure 4.1 Furuno integrated bridge system, electronic chart system, combined ARPA/radar, electronic bridge notebook, central consol with engine room/bridge control, interfaces from log and speed indicators, echo sounder and satellite communication units, steering gear and the thruster unit controls, alarm monitoring for off course/malfunction and feedback from propellers/rudder(s) etc. and position monitoring systems (GPS, DGPS, and Loran)

vessels to comply with the requirements at least two qualified operators (persons holding GMDSS certificates) have to be carried as part of the ship's complement. Additionally, to conform with the Standards of Training, Certification and Watchkeeping (STCW '95) convention, it will be a requirement for all ship's officers to be trained and certificated to GMDSS standards in order to retain a valid Certificate of Competency.

Current practice is such that with wireless telegraphy being committed to the history books and satellite communications developing at such an explosive rate the Navigation Watch Officer has taken on the dual role of Communications Officer alongside his other duties. This can generally sit well on a vessel where manpower is not in short demand, but any vessel can come under pressure at any time, which could make the added communications duty a burden on an Officer of the Watch (OOW) who may already be overworked with safe navigational duties.

NB. A vessel involved in a SAR operation would be expected to establish a designated 'Communications Officer'.

The GMDSS system is designed in such a manner that the touch of a button automatically sends out a designated distress message together with a GPS position by means of Digital Selective Calling (DSC).

Ships will be fitted with new communications equipment having four functions:

1. Distress alert and reception.
2. All ships urgency and safety.
3. Routine ship to ship and ship to shore calling.
4. Routine ship to group calling.

Channel 16 will remain as the calling channel for non-GMDSS users, but fewer and fewer persons will be listening out on Channel 16.

Masters should note that the ship's existing radio licence allows full GMDSS use but personnel are expected to pass through a training course and become certificated operators for GMDSS communications. Each vessel will be issued with a nine-digit number known as the Maritime Mobile Service Identity (MMSI) for conducting radio traffic. (The MMSI is obtained from the licensing authority.)

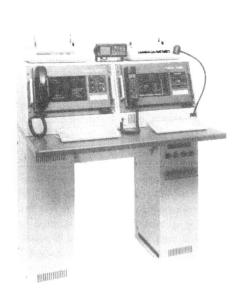

Figure 4.2 Standard radio GMDSS console including MF/HF radio station, DSC telex printer, VHF/DSC controller, Imarsat telex printer and Imarsat C terminal. (Manufactured by Raytheon Marine)

Requirements for the carriage of radio equipment

Sea Area A1 vessels will carry VHF equipment and either a satellite EPIRB or a VHF EPIRB. (English Channel – GMDSS A1 status effective 1 January 1998). *Sea Area A2* vessels will carry VHF and MF equipment and a satellite EPIRB. *Sea Area A3* vessels will carry VHF, MF, a satellite EPIRB and either HF or satellite communication equipment (Sat-Comms). *Sea Area A4* vessels will carry VHF, MF and HF equipment and a satellite EPIRB. Additionally, all ships will carry equipment to receive MSI broadcasts.

Ship identification transponders

The IMO has recently prepared new carriage regulations which require ships to carry identification transponders from 1999 onwards. Aircraft have carried such transponders for some time and in the case of war planes they were able to discern between friend or foe. The principle is the same, in that ships will carry a GPS, interfaced with radio equipment (GMDSS and GPS are both mandatory from 1999).

A shore station transmits an interrogation message on Digital Selective Calling (DSC) containing the co-ordinates of a geographic area. Vessels inside the area

respond, while vessels outside the area do not. Response by a vessel includes an individual nine-digit ship identification number known as a Maritime Mobile Service Identity (MMSI). This permits further communication, by voice or by obtaining additional information from the transponder.

Up to four information items could be requested by each interrogating message and may include:

Sensor information	Operator information
Position	Port of departure
Time	Destination
Course	Status
Speed	Cargo
Draught	Hazard category

Ship to ship identification

Ship to ship identification is possible by employing VHF, DSC, transponders and making the geographic area sufficiently small to allow identification of one ship by another. The MMSI identifies respective vessels and allows voice transmission, say in collision avoidance situations, with the confidence that it is clear which ship is being talked to. Increased technology will extend the use of and permit transponders to operate directly from the radar (ARPA) and the electronic chart (ECDIS).

Global radio transponders

Automatic Identification System (AIS)

The AIS is expected to become a mandatory carriage requirement by 2002 on all new tonnage over 300 GT and all passenger ships. At the time of writing it is included in the draft revision of Chapter V of the SOLAS convention and from current reports it is being considered as superior to the DSC operation undergoing trials.

The objective of AIS is to make any vessel equipped with a radio transponder visible to all other shipping by means of navigational data exchange. The items for continuous transmission could include: the ship's identity, its position as determined by on-board position sensors, its course and speed, the heading of the vessel and its rate of turn.

The concept is such that this information is available to all vessels on a continuous automatic transmitting basis and is expected to improve collision avoidance. The exchange is via a VHF data link and as such is 'on air', this in itself allows local VTS centres to receive similar data provided they were fitted with a base station.

The AIS is independent of radar, and because data is received automatically the benefits of reduced workloads on VTS operators cannot be overestimated, especially in busy shipping areas. Clearly with many different manufacturers involved a global standard to exchange information must be established and it is essential to have a single acceptable AIS for all.

Application of an on-board AIS must be seen to be an autonomous method of improving collision avoidance. It provides traffic information to the VTS operator while being independent from and additional to radar. It could also act as a ship reporting/monitoring system. Benefits from this system could be the reduced costs of radar equipment but still allow vessel monitoring through waterways like canals and rivers, which are without full radar coverage.

The most experienced mariners would accept the benefits of ship to shore transmission especially if it provides automatic operation. However, once ship to ship communication is involved, the past dangers of the misuse of VHF become a realistic concern. Station identification has always been the unknown and often dangerous element when combining the use of COLREGS and manoeuvres, especially in poor visibility.

recognized, certainly in the early days, that not all ships will be fitted with the AIS transponder.

The fact that the transmitting vessel will be aware of the data output is one aspect; the possibility of incorrect data being transmitted is another. This could have extremely dangerous and confusing results on the bridge of a receiving vessel.

Not all craft are over 300 GT, for example small fishing boats, leisure craft, etc., found in coastal regions, and would not be required to carry the transponder. Even larger fully equipped vessels cannot be guaranteed to have the system fully operational.

Collision avoidance methods have been considerably improved over the years with the introduction of traffic separation schemes, amendments to the COLREGS, and the involvement of VTS to name but three. The AIS should be as welcome as any of these, but used in conjunction with the many other aspects of seamanship. The fisherman doesn't expect to get his feet wet because of the advances in the 'big ship' technology.

Figures 4.3 and 4.4 show the need for effective communications.

Figure 4.3 Example of a weather satellite image over the North Atlantic region

At present ships operate plotting methods by either manual or automatic aids and maximize the information on a target inclusive of its relative motion. The AIS, if carried, would not provide this element. It must also be

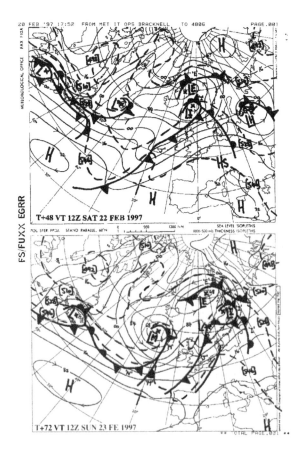

Figure 4.4

Radar: additional use

Major changes in basic radar technology generated the various display options ARPA (see Figure 4.5), Automatic Ship Identification, parallel indexing and advances in collision avoidance methods. As probably the most utilized of all navigation instruments it is difficult to understand why the shipping industry continues to have accidents related to the use of radar. The fact that it is human nature to make mistakes, through distractions or employing standard techniques in unusual circumstances or even taking short cuts in procedures, doesn't deny the fact mistakes do occur.

Figure 4.5 Nucleus 6000A Automatic Radar Plotting Aid (ARPA) mounted display by Kelvin Hughes set into integrated bridge

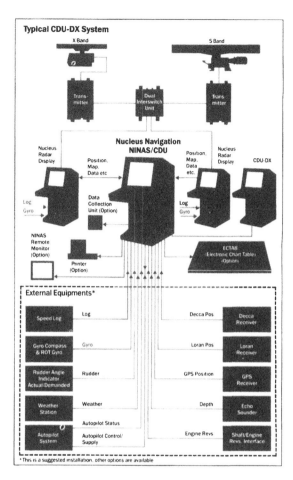

Figure 4.6 Interfaces of the Kelvin Hughes Nucleus Integrated Navigation System (NINAS)

Radar as an aid, when used correctly, is beneficial, but when engaged as an ARPA will still incur a delay while the data is assimilated, a delay that is not present with a line of sight situation. The trade-off being that line of sight may present a visible aspect of the target, where as the ARPA data provides enhanced information on the target once it is 'acquired'.

Figure 4.6 shows the interfaces of the Nucleus Integrated Navigation System (NINAS) and Figure 4.7 shows examples of two free-standing GPS chart top models.

ECDIS

The Electronic Chart Display and Information System (ECDIS) has been jointly developed by the International Maritime Organization (IMO) and the International Hydrographic Organization (IHO). The Electronic Navigation Chart (ENC) of this system contains all the charted information considered necessary for the safe navigational procedures exercised by the mariner and may in addition contain 'sailing directions', not necessarily found on paper charts.

The ECDIS satisfies the legal requirements for a vessel to carry charts, and provided the vessel is equipped with capable and reliable display instrumentation, and a back-up arrangement is retained on board, she will comply with Regulation V/20, of the 1974 SOLAS convention. System qualification would also require that a position sensing device is incorporated, together with an approved correction/updating service.

The Admiralty Raster Chart Service (ARCS)

The ARCS is a digital reproduction of the Admiralty Paper Chart which is capable of being employed by many types of electronic navigation systems, including ECDIS. Essential support to ARCS is provided by a comprehensive correction service which mirrors the

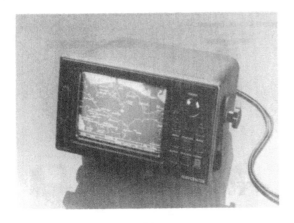

(a)

(b)

Figure 4.7 Free-standing GPS chart top models

existing Weekly Notices to Mariners but is issued as a weekly compact disc (CD).

Owners and users of ARCS with compatible operating equipment have the choice of either one or two service levels:

1. ARCS-Navigator, which provides the latest updating information under licence, inclusive of regional CDs and the weekly correction CD service. Additional charts (CDs) can be added to the initial package as required, the user paying only for what regions are needed.
2. ARCS-Skipper, a package for an operator which has less need for frequent updates. The user has access to charts but without the automatic updating service. The system is under licence without a time limit and the user can choose when to update. Some systems may incorporate a manual update facility which allows additional information to be inset onto ARCS from the Weekly Notices to Mariners.

ARCS charts are available in the form of a CD-ROM and provide worldwide coverage over 10 CDs with an additional CD for small-scale charts. Masters should be aware that there are many Electronic Chart Systems (ECSs) currently available which do not meet the specifications required by the SOLAS regulations and must only be used in conjunction with approved paper charts.

Advantages of approved Electronic Navigation Chart (ENC) systems

1. ENCs do not wear out and the cost of replacing and renewing is eliminated.
2. Once on board, the vessel is no longer dependent on chart suppliers to provide paper chart replacements.
3. No storage problems with software as compared with large drawers for chart folios.
4. Automatic chart correction from monthly CDs. This reduces the risk of human error in chart corrections.
5. Time saved in chart correction is cost effective and labour saving.
6. RCDS (Raster Chart Display System) provides a reliable and consistent correction service.
7. Full navigation information is available at a single control position, without having to move to a separate chart room/table.
8. Instant 'real-time' display of current position.
9. Position is shown relative to chart features and is immediate and without the distraction of requiring separate plotting activity.
10. They eliminate the need to check chart datum and then apply manual updates to the GPS.
11. Simplifies watch handover procedures.
12. Eliminates the need to change charts in busy waters.
13. Automatic input of courses eliminates human error risks in 'passage planning'.
14. Advance easy check for on-route hazards by just scrolling ahead on advance tracks.
15. Less human error in formulating passage plans where courses and distances are calculated automatically.
16. Overall workload and fatigue are reduced.

Working with ECDIS

The use of ECDIS requires the mariner to realize the capabilities and the limitations of the system over a period of sustained use. It should be understood, that it is not a replacement for a paper chart but more an integrated navigation system – a system where the Master can take a 'bird's-eye view' and see the response of his own vessel to direct orders and movement of the vessel's own manoeuvring aids.

Such a system will only perform effectively if the user is adequately trained. Activities such as passage planning, pilotage, manoeuvring and berthing will only enhance ECDIS to the end user, and then only over a period of familiarization.

Masters are advised to be fully aware of increased technology and the computer literate, 'hotshot, junior officer' who is aware of the fact that such a system will provide automatic functions to what was previously carried out manually. Not only will more detailed information be available from such equipment but it will be delivered and acted upon at a far faster rate, with more accuracy than was previously possible.

The obvious danger of such a system is that over-reliance can take place and in the event of failure, or false information being displayed, it would become the responsibilities of the courts to determine where the blame would lie in the event of an accident. To date, this situation has as yet not been tested. The reliability of hardware and software within ECDIS operations, manufactured under quality control conditions, still gives no guarantee against a power failure or component malfunction.

Back-up systems, to assume rapid takeover in the event of a mishap, especially in a critical situation, would seem to be the only way to retain the confidence of navigators. A system failure, on approach to a harbour entrance in poor visibility, for instance, could destroy months of confident familiarity with a system. In such a scenario, however, paper charts would not have been an adequate substitute and could not be seen as an effective back-up.

The final product will be judged by the person using the system. It will be the user who decides on what is displayed to conduct the task in hand and it will ultimately fall to the mariners to manage and master the equipment for both open and inshore waters.

Position fixing

Paper charts

One of the benefits of having a bridge team in position is that minor mistakes by individuals can often be picked up and corrected by other members within the team. This is especially so when radars are operational with alternative primary and secondary position fixing systems working alongside each other. The use of 'parallel indexing', providing continuous on-track monitoring, should also be considered an essential element of radar use and position monitoring, so reducing errors and maintaining the margins of safety. If the vessel's navigational passage is well planned it should incorporate radar conspicuous targets and identified areas where parallel index techniques can be employed.

Visual fixing

Masters should ensure Watch Officers engage in visual fixing with a minimum of three position lines in coastal regions, as a primary position fixing method. Watch Officers should be cautioned against the use of floating marks and preferably employ fixed land marks for obtaining bearings. Visual fixing should be practised in conjunction with instruments to enable one method to be cross-referenced with an alternative.

Instruments clearly have an essential place both as primary and secondary methods and are obviously required extensively in poor visibility. However, in the modern day of GPS, instruments are generally not too restricted by range, but their use could be deemed a distraction and detract from the 'lookout duty' by the OOW. It should be noted that many more accidents happen in coastal waters than in the open waters at extended range and well away from land, and this would reflect poor monitoring of the ship's position by whatever means in coastal regions, e.g. the grounding of the *Exxon Valdez* in 1989 off the Alaskan coast brought into question standards of training and the issue of VTS operations and the grounding of the *Sally Albatros* in 1994 in the Baltic Sea, in ice conditions, questioned the position monitoring accuracy of the vessel.

A suggested format of Master's standing orders for position fixing should generally include the following:

1. The OOW should employ both a primary and secondary position fixing method when in charge of the navigational watch.
2. Visual fixes should be gainfully used employing a minimum of three position lines.
3. In poor visibility alternative position fixing instruments must be used.
4. Floating buoys and beacons must not be used for position fixing, and if making reference to such navigation marks, the position of these markers must be cross-referenced on the chart.
5. Position fixing must be established at regular intervals to suit the navigational circumstances to enable full analysis to be made of the vessel's progress, i.e. coastal fixes would be expected at a greater frequency than deep sea positions.
6. All positions, when applicable, should be correlated with soundings.
7. The Master must be informed if position fixes indicate that the vessel is off the designated track.
8. Any malfunction in compasses or position fixing instruments must be reported to the Master immediately.
9. The transfer of position from one chart to another must be checked by latitude and longitude as well as by bearing methods. Due regard should be taken

account of the scale(s) of the two charts and the respective data being employed.

10. Should the vessel's position give rise for concern at any time, the Master must be informed as soon as possible.

Additionally many shipping companies expect the vessel's progress to be monitored by parallel index methods or alternative mapping techniques.

Electronic charts

Where electronic chart systems are in use 'way points' should be employed to give adequate clearance from hazards, shoals, landmarks, etc. and tracks should take account of prevailing weather directions. The ship's progress should be monitored on a regular basis at intervals that depend on the circumstances, although electronic chart systems are interfaced with a GPS or DGPS which provides virtually continuous position plots. Overreliance on electronic position fixing methods can become the navigator's Achilles' heel if instrument failure or irregular power supply occur.

On 10 June 1995, the passenger vessel *Royal Majesty* grounded on the Rose and Crown shoal approximately 10 miles east of Nantucket Island, Massachusetts. The cause of the incident was found to be that the GPS antenna cable had become partly disconnected which caused the GPS to switch to 'dead reckoning mode'. This situation went unnoticed and the autopilot continued to respond to the information derived from the GPS. The set of the vessel, caused by wind, current and sea conditions, remained undetected by the Watch Officers for 34 hours prior to the grounding.

For Masters and Watch Officers alike the lessons to be learned from the *Royal Majesty* incident may include any or all of the following points:

(a) Overreliance by watchkeepers on automated navigation features within the bridge system and overall poor navigational practices in general.
(b) No check on the primary method of position fixing by an independent source.
(c) Echo sounder alarm being previously set to zero depth effectively renders underkeel clearance alarm systems null and void.
(d) Remoteness of the GPS receiver, which had a short duration audible alarm and sounded when switching to dead reckoning mode, contributed to watchkeeper's not noticing any change occurring.
(e) When making a landfall, additional practices and special care should be exercised.
(f) Integrated navigation systems require adequate training. (NB. New systems may require operators to be updated.)

Parallel indexing

Principle

The basic principle of monitoring a ship's movement by means of parallel indexing methods can be realized by noting that when a ship is passing any fixed object, and steaming on a steady course, that fixed object can be observed to move in a reciprocal direction at the same speed as the vessel.

If observing on radar, in the 'relative motion' mode, all fixed objects would appear to move across the screen in the opposite direction.

Example 1

Assume a vessel to be on a course of 260° (T) heading down the English Channel towards the Lizard at a speed of approximately 12 knots, expecting to pass 5 miles off Lizard Point.

The charted positions at 30′ intervals could expect to appear as indicated in Figure 4.8.

If monitoring on radar at the same time (relative motion), then with the ship's position seemingly fixed into the centre of the screen, the Lizard Point would appear to move in the opposite direction, as indicated in the figure by the pecked line A–B.

On the radar screen this movement would appear as indicated in the Rad-Scan of Figure 4.9.

Practical application

In practice if the reflector plotter is employed using the Variable Range Marker (VRM) and the parallel bearing cursor the above can be demonstrated.

The bearing cursor is aligned with the desired 'course to be made good', and the VRM is set at the distance off, in this case 5 miles. Use a Chinagraph pencil and a plotting ruler to draw in the reciprocal course from the fixed radar target (i.e. Lizard Point) as indicated in the Rad-Scan of Figure 4.10

Summation

It would follow that, provided the ship maintained its course, the Lizard Point would also maintain this reciprocal movement across the screen.

When engaged in passage planning, if the navigator plotted the reciprocal course from the Lizard, then as the ship proceeded past the Lizard, any deviation by the ship from the course track would result in the echo of the Lizard Point moving away from the designated (pecked) plotted line.

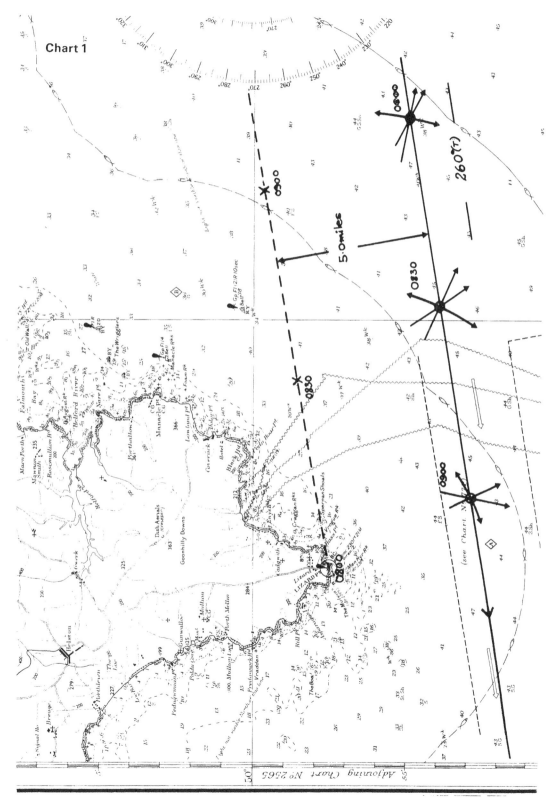

Figure 4.8

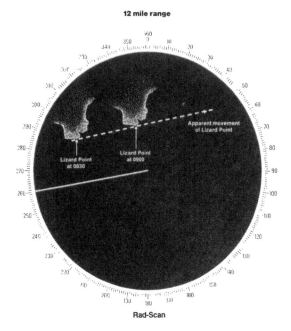

12 mile range

Apparent movement
of Lizard Point

Lizard Point
at 0830

Lizard Point
at 0900

Rad-Scan

Figure 4.9

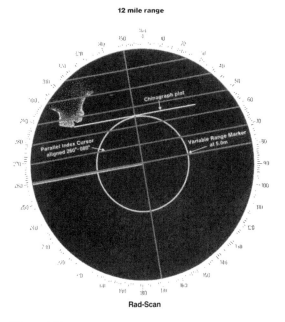

12 mile range

Chimagraph plot

Parallel Index Cursor
aligned 260°–080°

Variable Range Marker
at 5.0m

Rad-Scan

Figure 4.10

Deviation by vessel

If the same ship is caused to set towards the coast by currents or traffic avoidance action, she will experience some course other than 260° (T) (Figure 4.11).

When engaged in parallel index plotting the results of this deviation would be clearly indicated on the radar

screen as shown in the Rad-Scan of Figure 4.12. The Lizard Point would be seen to move inside the plot line A–B, indicating that the ship has been set closer inshore than the desired 5 mile distance off required.

The amount the ship has been inset can be obtained by employing the VRM to measure the distance against the parallel index cursor line, now setting through the Lizard Point. (Inset approx 1 mile – assuming parallel bearing cursor lines at 2 miles distance apart.)

Parallel index plotting of this nature can allow continuous monitoring of the vessel's track and would indicate the direction and amount of any deviation from the desired course.

NB. Seamanship has many definitions. To some it could be the successful transportation of cargo from one part of the world to another, while others may perceive it as the avoidance of collision between ships. Whatever form the definition of seamanship may take, position monitoring and the keeping of an effective lookout continue to be rated as high desirables when considering this vast topic.

In the author's opinion parallel index techniques may lend more to navigation, but position monitoring and watchkeeping duties must be considered an aspect of seamanship no professional mariner would wish to contain under a single heading.

Course alterations

Using the basic principles described in Example 1, alterations of course and the vessel's progress on track can be tightly monitored from one heading to another.

Example 2

A vessel is navigating around Land's End and is currently steering 331° (T) and plans to alter course at a position south west of Longships Light.

NB. The Longships Lighthouse is a prominent radar target and as such is ideal for employing parallel index methods.

With reference to Figure 4.13 the desired tracks are 331° (T) and 006° (T) at distances of 3.0 miles and 2.1 miles respectively.

Practical application

The plotting on the reflector plotter is carried out in two steps.

1. Align the parallel bearing cursor on 331° and set the VRM to 3.0 miles. Using the plotting ruler and chinagraph pencil mark off the parallel index line of 331°/151° back from the Longships Lighthouse.

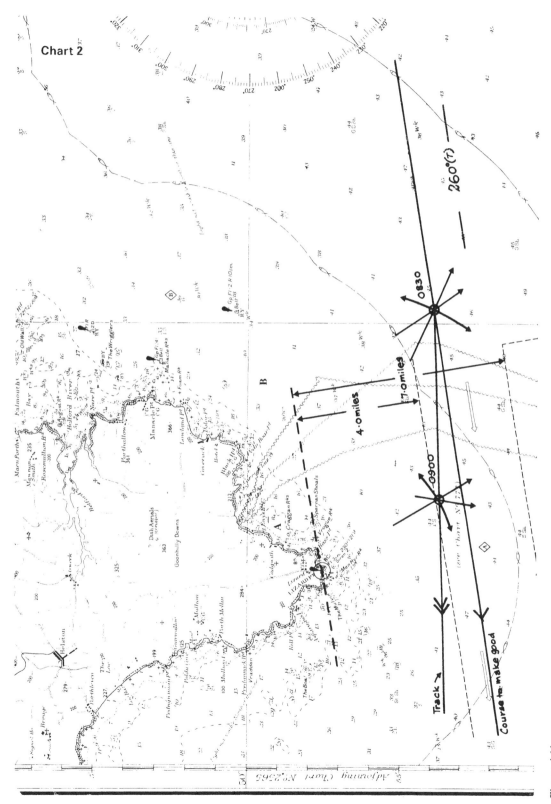

Figure 4.11

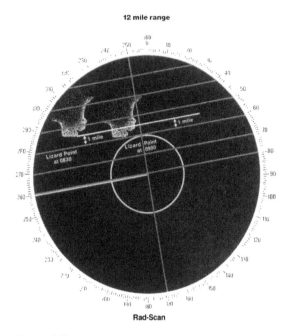

12 mile range

Rad-Scan

Figure 4.12

2. Align the parallel bearing cursor on 006°/186° and set the VRM at 2.1 miles. Construct the parallel index (second) line to represent the second course.

Summation

The planned alteration of course position is then represented by the intersection of the two constructed index lines.

Provided the Longships Light moves down the first index line of the plot, the observing officer is confident that the vessel is approaching the alteration position as planned (see the Rad-Scan of Figure 4.14).

NB. A wheelover point 0.5 miles ahead of the alteration point has been shown to indicate the time to execute the manoeuvre. Such a point would lie on line A–A, on Figure 4.14, and would hold good irrespective of whether the ship is on track or not.

Line B–B, represents a 'margin of safety' and can be introduced to the plot as and when appropriate. In this example it might be considered as a limit of any deviation to starboard (see the Rad-Scan of Figure 4.15).

Precautions in the conduct of parallel indexing

Parallel indexing can be used in conjunction with several aspects of navigation, including position monitoring off a coastline, holding station within a traffic separation

scheme, when engaged in search and rescue activity, or maintaining navigational accuracy inside a narrow channel within preset margins of safety. However it is used, it is suggested that Masters should encourage indexing in clear weather as well as in poor visibility as a means of on-board training by Watch Officers. It should also be stipulated that the use of indexing does not eliminate the use and need of other position fixing methods and that effective watchkeeping must include both a primary and secondary position fixing method wherever possible.

It is imperative that in proposing and conducting parallel indexing, the 'index mark' is correctly identified on the display, from the onset of the exercise. Failure to confirm this 'mark' could result in an index-assisted grounding situation, for which vessels would be especially vulnerable in poor visibility.

It should also be remembered that an appropriate range scale is employed on the radar and that when index lines are drawn on the reflector plotter they are good only for that specific range scale. Use of index lines on an alternative range is useless and must be considered dangerous. Also the tracks of the passage plan should incorporate appropriate indexing lines and these need to be confirmed as safe or indexing could become a recipe for disaster. In the event that the vessel is seen to move off track, immediate action is required to rectify the situation and in some cases may generate the need to conduct an alternative parallel indexing.

The observing officer, when intending to engage with indexing techniques, should always ensure that the radar performance is satisfactory and that the accessories for use with the reflector plotter are available and adequate for the task. Distractions halfway through an indexing operation are to be avoided and navigators need to concentrate from the start to the completion of an indexed manoeuvre.

For additional references on the topic of parallel indexing, see Merchant Shipping Notice M1158, or *Parallel Index Techniques* by I. Smith and R. Mulroney, published by the Stanford Maritime Press.

Fatigue

The employment of watchkeepers

Much has been written and debated over the last decade on the subject of fatigue affecting personnel at sea. The fact that people do become tired and make more mistakes than the norm after long hours of duty is well known. The subject is probably one of the most discussed issues of our time, and the reasons why it is debated to such lengths are money and the desire for an ideal world.

The reader should be well aware that the ideal world is the ideal dream, and the Mercantile Marine is a long way

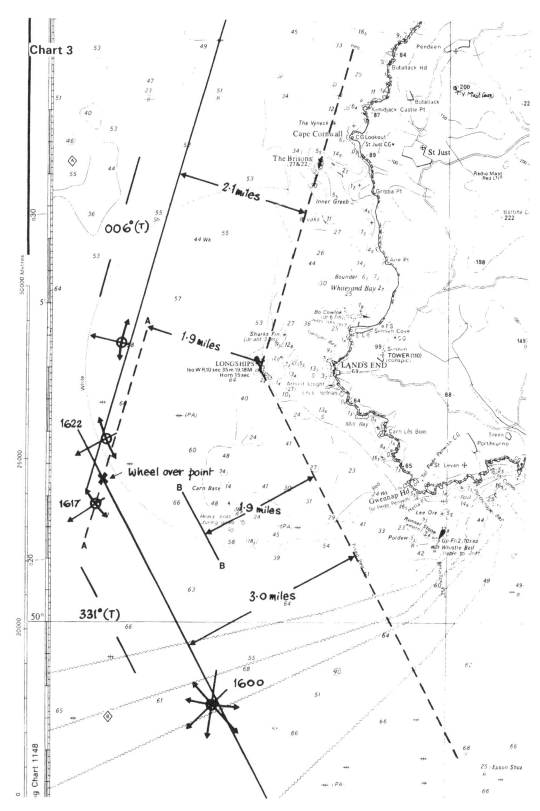

Figure 4.13

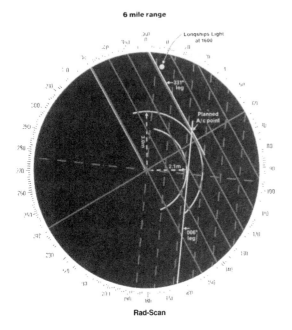

Figure 4.14

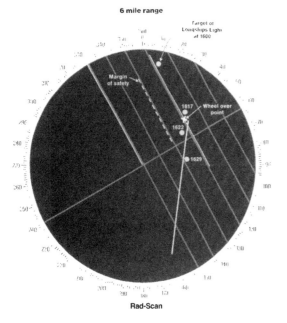

Figure 4.15

from achieving such a level. The marine environment is a practical one and as such is influenced by the factors of available manpower, the qualities of that manpower, and most influential of all, the economic factors affecting the operation.

No individual, least of all a shipping company, would willingly agree to move down the path of insolvency,

liquidation, and bankruptcy for want of a few choice words. To put an additional watchkeeping officer on every ship may tackle the question of fatigue, but is the manpower available to do this, and are the additional personnel going to possess the necessary qualifications. (This raises the question of training budgets and do all companies actively participate in training.)

Would the ships need the additional personnel at all times and even if they were available to act as relief officers, could the shipping company pay the cost of wages? Would all companies agree to take on more personnel to relieve the fatigue question and would they also pay the additional training costs? Would existing seafarers accept a pay reduction to cover the costs of employing additional manpower?

Once these questions are asked the reality of not living in an ideal world tends to come home to everybody. The ship's Master would welcome the luxury of an additional watchkeeper, both deck and engine room, and nobody needs them more than when, for example, arriving at the Dover Straits in fog and realizing the need for employing double watches.

The future is under even more scrutiny if single-man watchkeeping is in operation, as was seen with the *CITA* incident which grounded on the Scilly Isles in March 1997. The report read that 'the incident clearly demonstrates the risk of operating ships at night with inadequate numbers on watch on the bridge'.

The fact that the ship was operated in UK waters in contravention of UK watchkeeping requirements, where an OOW fell asleep causing such a disaster, is of direct concern to all.

A Master's concern

In October 1998 the Norwegian vessel *Stina* struck the island of Kerrara in the Firth of Lorne. The MCA prosecuted the Master who admitted falling asleep up to 45 minutes before the ship struck rocks at full speed.

The Master had an unblemished record for over 27 years, and this was taken into account when he was fined £1000.

The above incident was also noticeable because of the risk of operating with only a single person on the bridge. The dangers this represents became a threat not only to the crew of the vessel but also for the environment and for other persons in the vicinity.

Combating fatigue

If ships are continually allowed to operate with reduced levels of manning because of improved automation, the back-up systems provided on board must be seen to be improved. Safeguards need to be incorporated at every stage of each operation to eliminate the risk of accident. Reports on past incidents have revealed serious flaws in the management of watchkeeping arrangements and this alone would fall directly onto the Master's shoulders.

Masters are not the most popular of people when they decide to double up watchkeeping personnel. In this situation Masters should not be concerned with popularity, their prime aim is not to risk the lives of the rest of their crew and other seafarers. The prudent decision to double up needs to be made with thought. If it is made too early and maintained, officers will get tired quickly. If it is made too late, or not at all, there may be disasterous consequences.

STCW provision

In accordance with the watchkeeping provisions as stipulated by the STCW convention, flag states will have to enforce mandatory, minimum rest periods to prevent fatigue and ensure fitness for duty. All persons who are assigned duty as an officer in charge of a watch, or as a rating forming part of a watch, shall be provided with a minimum of 10 hours' rest in any 24-hour period.

This period may be divided into no more than two periods, but one of these periods must be at least 6 hours in length. Notwithstanding this provision, the 10-hour rest period may be reduced to not less than 6 consecutive hours, provided that any such reduction shall not extend beyond 2 days and not less than 70 hours' rest are provided in each 7-day period.

Pilotage and the marine pilot

The local knowledge and skills of the marine pilot have long been an accepted necessity for the ship when entering and proceeding in coastal waters. The need for the services of the pilot have in no way diminished and if anything have increased with the much larger, deeper draught vessel becoming more predominant. The location of the port will reflect whether a sea pilot is employed before a river pilot and whether the Pilotage Authority would expect the use of a docking pilot. In any event the Master is expected to be fully conversant with the change from one type of pilot to another and be aware of the respective needs of each.

Boarding pilots: Master's duties and responsibilities

It is usual practice for the Master to have the 'con' of the vessel when approaching any pilotage station.

This task can be made considerably easier with advance planning, which if carried out in detail for each occasion will generate a smooth, incident-free engagement. The prudent Master would normally familiarize himself with the boarding area and establish an approach plan after initial communications with the station and reference to the chart and sailing directions.

Pilotage rendezvous/engagement area

The boarding position for the marine pilot is identified on the chart and usually detailed in the Admiralty Sailing Directions of the area. However, the Master should ensure that the position has adequate underkeel clearance for the vessel at its existing draught. It should also preferably be clear of strong currents and tidal effects at the time of engagement (reference ship's ETA to Admiralty Tide Tables/Tidal Stream Atlas, etc.).

Surface obstructions such as buoys or piers should be clear of any swinging room, which the Master may have to employ to create a lee for the pilot boat in relation to the prevailing weather direction. Relevant sea room, away from other traffic or harbour operations, would be the ideal, but this is not always possible to achieve.

If anchors are to be used a designated anchorage site with no underwater obstructions and good holding ground would be a useful location as a fall-back contingency position, if available.

Figures 4.16 to 4.19 show examples of typical pilotage situations.

Figure 4.16 Under pilot's control. Roll on-roll off ferry negotiates a buoyed channel in close proximity to the shoreline

Pilotage approach plan for pilot boat engagement

1. Plot inward and outward tracks to and from the pilotage station, which will provide an adequate *underkeel clearance (UKC)*.
2. Observe and identify any *'NO GO AREAS'* near the charted tracks.

Figure 4.17   Under pilotage. A passenger/vehicle ferry departs the French Port of Calais on course for passing through the breakwater entrance

Figure 4.18   Under pilot's control. Roll on-roll off/passenger ferry negotiates a turn in the close confines of Heysham Harbour, prior to berthing port side

3. Note on the chart any *communication positions* between the ship/pilot station.
4. Consider when *engines need to be placed on standby* and where the approach speed needs to be reduced and what the anticipated engagement speed is to be. (Bear in mind the hazards which might be encountered through interaction.)
5. Establish which *position fixing methods* are to be used on approach, employing both a primary and secondary system.
6. Identify the *least depth* and areas where the echo sounder would need continuous monitoring regarding the underkeel clearance.

Figure 4.19   Under pilotage. Large tanker with tug in attendance in the River Mersey, off Liverpool

7. Establish on the chart when the *anchors need to be cleared away* and when personnel would be expected to standby anchors for emergency use.
8. Note a *contingency* safe anchorage, in the event of the plan not proceeding.
9. Clarify that no *navigational hazards* have been recently issued for the pilotage area.
10. Check *communication* channels/frequencies are as scheduled with the Admiralty List of Radio Signals and are compatible with *local by-laws* or other port/harbour regulations.
11. Ensure that *adequate personnel* are available to carry out all operations envisaged with embarking a pilot, e.g. manual steering, rigging ladder, meeting pilot, etc.
12. Note and establish when the *bridge team* must be *in situ* to conduct a safe engagement. (Relevant to when the Master must assume the 'con'.)

NB. Once the pilot has embarked he becomes an integral member of the bridge team.

Types of pilots and pilotage

There are many varieties of marine pilots to suit the many differing types of pilotage and consequently the Master's approach and formalities will need to differ to match the circumstances of the engagement. To exemplify this, a sea pilot probably may not be influenced as much by the ship's draught and underkeel clearance as, say, the river pilot would need to be.

Specialized craft may also require special pilots, for VLCCs or air cushion vessels, for example. Other geographic needs could well dictate the need for speciality, in the form of estuary pilots, Grand Barrier Reef – passage pilots, ice pilots or the distinctive docking pilot. Some shipping companies that operate regularly out of one port would often employ a single pilot for all their vessels.

In general, marine pilots are drawn from a maritime background with extensive local knowledge of potential navigational hazards, port/harbour communications, local by-laws, and a level of expertise in ship handling. They must be physically fit in order to board the vessel in the first place and once aboard would need a degree of 'power of command' to ensure a confident approach to the task ahead.

The Master/pilot relationship

So much has been written and debated about the relationships between ships' Masters and marine pilots that the subject would fill a book in its own right. The Master/pilot relationship is of one man trusting another with his ship, his own livelihood and not least the lives of his crew, albeit for a comparatively short period of time.

However, the efficiency of the pilot station reflects on the individual pilots. Are the communications comprehensive? Is the pilot punctual at the rendezvous? Are the pilot boat crew seamanlike in their duties? These points undoubtedly influence a preconceived idea of the type of person about to 'con' the vessel.

Equally the actions of the Master are being analysed, but in reverse by the pilot. Will the Master ask the usual questions regarding ETA berth, use of tugs or anchors, navigational hazards, shoals and so on? How efficient is the Master, and how tight does he run his ship?

Generally there are two levels to consider in the Master/pilot relationship: a personal, man-to-man level, and the ever important professional level on a business footing. The pilot will become an additional member of the bridge team whose expertise should relieve anxiety and contribute positively to the safe control of the ship. It is not the pilot's role to take command of the vessel and this impression should not be given; the ship's Master retains the command authority in virtually every case, and in the event of a dangerous situation developing he must act to override the pilot.

Master/pilot: exchange of information

Once the pilot has boarded, the Master would inspect the pilot's authority letter and add the ship's general particulars to the relevant documents. The Master should inform the pilot about the immediate condition of the ship with regard to speed, course, position and draught and indicate the operational details regarding manoeuvring revolutions and relevant speeds.

Additional detail such as the general particulars of the vessel are commonly displayed on the bridge in the form of a pilotage card. The compass heading, being a variable, is normally noted in the proximity of the helmsman's position on a compass heading board and any compass error should also be displayed and the pilot advised accordingly.

Consultation between Master and pilot must take account of the proposed passage plan (see MGM '72). Bearing in mind that passage plans are expected to cover a berth-to-berth route and be flexible according to the circumstances, any amendments to the plan should be mutually agreed between the two parties.

NB. Because passage plans cover the pilotage area, it is debatable whether the pilots themselves should have a greater input into the final and commencing stages of a voyage. The present system, where the Master buys in the expertise of the pilot, would seem adequate provided the pilot's advice in adjusting the plan is accepted. It could also present logistical problems and cost more for a pilot to actively contribute towards the plan, when it would normally be constructed by the ship's navigator. Alternatively, the pilot could arrive on board with his own charts and his own pilotage mini passage plan, but a prudent Master would be reluctant to control his vessel via charts which may be of dubious quality and for which the Master would have no maintenance responsibility.

What the pilot brings to the ship is up-to-date local knowledge which reflects the current navigational detail of the area. He will be able to supply particular information regarding navigational hazards and the degree of underkeel clearance at various stages of the pilotage. As an active member of the bridge team his foreknowledge of communications will be essential for a trouble-free passage. As a docking pilot he would advise on berthing detail and the deployment of tugs, and satisfy obvious questions from the Master to ensure the effective operation of the vessel.

The interchange of information between two professional mariners will generate a level of mutual respect, where each can expect to gain from the other's knowledge, skills, and experiences.

Bridge team responsibility

It is recognized internationally that when the pilot boards and advises the Master regarding the control of the vessel's movements, the pilot is acting with the consent of the Master and as such the bridge team would be expected to respond positively towards the safe progress of the ship. The pilot, although an integral member of the bridge team and one who may well relieve anxiety, does not assume the command, nor does he relieve the

Watch Officer of his basic duties with regard to position monitoring and maintaining an effective lookout.

The command authority under pilotage

Historically it has never been the intention for the marine pilot to assume the command of the vessel, and on occasions where the vessel is in danger, or where the pilot exhibits a level of incompetence, then the ship's Master's command authority becomes essential. The Watch Officer, as the Master's representative, must therefore be fully aware of his total responsibility to act on behalf of the Master, in his absence.

Under such circumstances, where the safety of the vessel is brought into question, the Master, in not doing anything is not justified in the eyes of the law. He is obliged to act in the best interests for the safety of the vessel, passengers and crew.

NB. Fortunately, the need for the Master to override a pilot's actions is rare, because generally the pilotage standards and competence exercised are beyond question.

Duty of the pilot

The sole duty of the marine pilot is to conduct the navigation of the vessel, and he has no other power on board. It should be realized that the pilot is responsible for his own actions in the eyes of the law, while the Master's right to interfere is restricted to those circumstances where clear evidence of the ship's safety, or the inability or incompetence of the pilot, is observed.

Safety of the pilot

It has long been recognized that unlike a shoreside environment, the ship at sea, as a moving and unsteady platform, is exposed to greater risk. The approach of the pilot boat and the dangers of interaction during the boarding period could pose a particular hazard to the pilot. Bad weather and associated rough sea conditions can also clearly affect the safety and well-being of the pilot.

First, the pilot can expect a level of competence from the crew to contemplate stepping onto the ladder, observing the position of ladder and its physical condition, together with the level of preparedness by standby personnel. The assumption that a responsible officer will be designated to meet the pilot is very complimentary, but the need for the officer to be first and foremost a seaman is essential. Checks should reveal that the ladder is rigged correctly and the securing hitches are correctly fastened.

If an accident were to occur during boarding, the Master must answer the question, was an officer instructed to check the rigging and securing of the ladder prior to engagement? The danger to the pilot is not just the incompetence of the officer who secures the ladder, but overlaps the complacency of the responsible officer, and reflects on the Master not operating a tight ship.

The pilot boat is easily the best rescue craft, in the event of the pilot finding himself in a 'man overboard' situation. Ideally, it is already launched in position on site, and available for immediate response. It is to this end that the majority of pilot craft will land the pilot onto the ladder, then pull off while the pilot ascends the ladder. The pilot boat finally pulling away once the pilot has attained the comparative safety of the ship's deck.

Should the pilot craft not hold station until the pilot is on deck, the Master could well be forced to launch his own rescue boat to effect recovery. The very worst situations envisaged are bad weather, during the hours of darkness and where the pilot is thrust into a 'man overboard' scenario. The Master would be forced to react with communications, shiphandling, emergency rescue procedures and at the same time maintain effective navigational practice regarding position fixing, lookouts and underkeel clearance. No easy task for a full bridge team, even on a tightly run ship.

NB. Most pilot authorities are now equipping their pilot boats with increased manpower and means of recovering a person from the water in the horizontal position. New generation craft are often fitted with either House Recovery Nets, or mechanical stern platforms or other similar equipment to assist the recovery from the surface.

See Figure 4.20 for a Marine Guidance Note as to the manning of pilot boats.

Compulsory pilotage

Many areas of the world have established compulsory pilotage conditions and these are often employed on the basis that incoming ship's Masters cannot be expected to be familiar with every port or harbour. Other more specific conditions could well dictate the needs of compulsory pilotage and be any one, or several of, the following factors:

(a) The size of the vessel could influence the type of pilotage requirement.
(b) The volume of traffic within the harbour limits.
(c) The nature of cargo being carried by the vessel, e.g. hazardous cargoes.

MARINE SAFETY AGENCY

MARINE GUIDANCE NOTE

MGN 50 (M)

Manning of Pilot Boats

Notice to Competent Harbour Authorities, Pilots and Pilot Boat Crews

This Note replaces Merchant Shipping Notice No. M1473

1.0 Introduction

1.1 Following the death of a pilot in July 1990, the Marine Accident Investigation Branch (MAIB) recommended that all pilot boats should be manned with a minimum of two crew including the coxswain. The MAIB further recommended that the competent harbour authorities should ensure that all the crew members of their pilot launches attend a first aid course and that it should be the policy of these authorites that the crew should carry out man-overboard retrieval exercises at regular intervals. The Department of Transport accepted these recommendations and issued the following guidance which still stands:

1.2 Manning

1.2.1 Every pilot boat shall be manned by a minimum of two adult persons, namely a coxswain and a deckhand who can assist the pilot when boarding or landing. The competent harbour authority or owner of the boat shall be satisfied as to the competence and fitness for duty of these persons.

1.2.2 A second crew member on a pilot boat is essential for observing the pilot and the pilot ladder:

a) when the pilot boat comes alongside the ship;

b) at the time the pilot transfers between the pilot boat and the pilot ladder; and

c) when the pilot boat departs from alongside the ship.

A second crew member is vital if there is a need to recover the pilot from the water.

1.2.3 All pilot boat crew members shall-

(a) hold a Marine Safety Agency (MSA) First Aid at Sea certificate; or

(b) hold a First Aid certificate issued in accordance with Regulation 3(2) of the Health and Safety (First Aid) Regulations 1981 (SI.1981/917), or

(c) have received training in emergency first aid in accordance with Regulation 3(2) of the Health and Safety (First Aid) Regulations 1981 (SI.1981/917), as described in paragraphs 48 and 58 under Regulation 3(2) of the Health and Safety Commission publication "First aid at work - The Health and Safety (First Aid) Regulations 1981 - Approved Code of Practice and Guidance" (ISBN 0 7176 1050 0).

1.3 Man-Overboard Retrieval

1.3.1 Competent harbour authorities should require man-overboard retrieval exercises to be conducted by each pilot boat crew at intervals of not more than six months.

2.0 Further Recommendations

2.1 Following the death of a crew member of a pilot boat in June 1996, the MAIB recommended that the Marine Safety Agency review the adequacy of the man-overboard retrieval system carried on board the vessel.

2.2 The MSA, having reviewed the system, considers that the possibility of having to retrieve a crew member who has fallen overboard should be covered in the periodic exercises referred to in paragraph 1.3.1 above.

2.3 In addition, the retrieval procedure should be covered in the functional tests of the retrieval equipment demonstrated to the satisfaction of the Certifyung Authority which carries out the survey and issues the certificate for each pilot boat.

Where a vessel is normally manned by a helmsman and one crew member, the demonstration required by these functional tests should include retrieval of a crew member from the water. (In this demonstration, the crew member can be assumed to be conscious.) This demonstration should assume that the pilot boat is in the minimum manned condition ie with only the coxswain and deckhand on board, and that the deckhand falls overboard and has to be recovered.

Seafarers' Standards Branch
Marine Safety Agency
Spring Place
105 Commercial Road
Southampton SO15 1EG

Tel 01703 329242
Fax 01703 329252

December 1997

MS 7/8/1256

© Crown Copyright 1997

An executive agency of
**THE DEPARTMENT OF THE
ENVIRONMENT, TRANSPORT
AND THE REGIONS**

Safe Ships Clean Seas

Figure 4.20

(d) The anticipated level of difficulty in completing the pilotage.
(e) The duration of the pilotage.
(f) Any special manoeuvres that the vessel could be expected to carry out.
(g) Special factors which relate to shipping safety and efficient port operations, e.g. canal transits, ice conditions, etc.

In the event that a pilot could not fulfil his duties while engaged aboard a vessel inside a compulsory pilotage region, the options open to the Master would be somewhat limited. Each situation would have to be treated on its merits, but in general the Master would be expected to take the immediate 'con' of the vessel, and open up communication with the Pilotage Authority.

The first priority is to ensure the safety of the vessel and to this end the Master should attempt to hold station in the present position. This could prove difficult against currents and weather. One option would be to update the ship's position and obtain a full chart assessment with the view to taking the vessel to a contingency anchorage to await a new pilot.

Depending on the geography and the nearness of a suitable holding position, the Pilotage Authority could advise the Master to continue under caution to a more acceptable position and there hold for a relief pilot to board. In any event the Master would be reluctant to continue beyond such measures, other than to ensure the security of the vessel and would most certainly seek advice from the authority, prior to proceeding further.

Vessel traffic service (VTS)

Introduction

A vessel traffic service (VTS) is a system which is used to monitor the progress of ship's movements within the confines of harbours, ports and their approaches. They are extensively employed around the world's main shipping areas in busy shipping channels, rivers and canals with the view to providing navigational information to ship's Masters, reducing the risk of collision and generally expediting the turnaround time of commercial operations.

Operational VTS

Most VTS systems are actively engaged with communications, radar surveillance, and more recently with electronic chart systems. There is little evidence that a system will influence the immediate tactical decisions required by vessels in close proximity to each other,

but advanced planning can certainly avoid congestion at focal points.

The level of success which a system can experience will be very much in the hands of the quality and effectiveness of the operators within the system. The IMO provides guidelines and training for operators, but the needs of any system will be identified with specific Harbour Authorities. The services of sea pilots, river pilots and docking pilots are normally incorporated in the VTS operation and generally handle all communications for the boarding and disembarking of pilots.

Objectives of VTS

In accordance with the IMO guidelines, the prime objective of VTS is to provide a service designed to improve the safety and efficiency of maritime traffic, while at the same time provide protection to the environment. In broader terms the aim is to minimize loss or damage to society by protecting coastlines and local communities from maritime accidents.

It is also a requirement of a working system to act in support of allied activities, such as search and rescue operations or other ship reporting systems. The VTS authority should be seen to comply with the law of the land and not impede the through passage of vessels engaged in innocent passage towards home ports or in transit through coastal waters.

IMO guidelines for VTS

In accordance with the IMO guidelines a VTS provides three basic functions, namely (a) an information service, (b) traffic monitoring and (c) traffic organization.

Responsibility and liability fall on the state which is setting up and engaging the system. A vessel entering into a VTS area where a mandatory ship reporting system applies is expected to contact the shore-based authority without delay. The Master of such a vessel could also expect to receive advice or instructions with respect to traffic safety and advance procedure.

It is important to realize the difference between *advice* and *instruction*, and the influence that the VTS can make upon the ship's Master. A level of interaction between the Master of the vessel and the VTS operator must be anticipated and both parties are expected to work in harmony to the common goal of safe ship movements. The operator's advice generally directs the vessel into a line of action that satisfies the overall activity of the VTS area and generally avoid close quarters situations. Under special circumstances, with concern for the safety of traffic, the decision-making authority of the Master may be superseded by the VTS authority and 'instructions' would be issued, i.e. an order.

If instructions are issued, then compliance of the same is mandatory and any non-compliance could involve sanctions being imposed. It must be assumed that a vessel involved in an accident would probably be held totally responsible if it were shown that the Master had not made full use of any assistance or information made available from the VTS authority. By the same reasoning, if advice or instructions were issued without due care and thought by the VTS then the liability following an accident could be seen to go the other way.

Example VTS

The Canadian Coastguard operate a total Marine Communication and Traffic Service which covers the east/west and northerly coastlines. A co-operative VTS is maintained between the United States and the Canadian authorities.

As an example, vessels 20 metres or more, in bound for the Vancouver area crossing longitude 127° W, latitude 48° N or within 50 nautical miles of Vancouver Island, would contact 'Tofino Traffic' (VHF Channel 74). Participation is mandatory with Tofino, Seattle and Vancouver traffic inside the Canadian and US territorial waters. Participation to seaward, typically up to 60 miles offshore, is on a voluntary basis but vessels are strongly encouraged to participate to receive the full benefit of the VTS system.

Shore-based pilotage/radar advice

Instructions from a shore-based centre regarding pilotage/radar advice could be defined as an act of pilotage in a designated area, conducted by a licensed pilot of the area, from a remote position other than on board the vessel concerned. The legal standing for the activity is within the Pilotage Act, but the operation must:

- be in accord with any national legislation,
- have set limitations and border limits prescribed as per local pilotage regulations,
- have the agreement and co-operation of associated VTS centres regarding traffic movements,
- have effective radar and communication equipment in order to conduct safe operations.

Shore-based pilotage may be incorporated into VTS systems, but a level of integration between Masters, VTS operators, and shore-based pilots (SBPs) can be expected in the future. Clearly circumstances of traffic density, adverse weather patterns and unforeseen emergency could influence where and when such shore-based activity can be allowed to proceed. In any event, for the system to be conducted at all the continuous presence of an experienced pilot being available in the traffic control centre would be an essential element.

Where a pilot is on board the vessel, a shore-based radar pilot would act as a second, advising pilot. In the event that the vessel has no pilot on board, then an SBP would advise the ship-borne navigator (Master) probably with regard to, for example, positional information, speed reductions in fairways, navigation warnings and traffic situations. It would be considered an exception for the SBP to take over the full navigational conduct of the vessel and the system must be accepted as an additional, co-operational service.

Advantages of shore-based pilots

(a) Advising the bridge team after an accident.
(b) Continuous navigational support to a vessel in fog or other critical situation.
(c) Additional advice when seeking a contingency anchorage.
(d) Advice to a Master on approach to a pilotage station.
(e) Support to a Master and pilot on board the vessel when considering the proposed route and amendments to the passage plan.

VTS: the legal implications

The VTS has certain legal implications in the following areas.

Masters, officers and crew
The Master remains ultimately responsible for the safe conduct and passage of the vessel. Any misdemeanours could lead to flag state or Port State Controls imposing sanctions on the vessel.

The shipowner
With the many offshore registers, second registers, and flags of convenience, the shipowner can expect to be liable for any damage caused by the company's vessels.

The cargo owner
Although damage caused by cargo could be extensive, the general belief is that the cargo owners are usually innocent when the time comes to apportion blame following an accident.

The underwriter

An underwriter is usually the most directly concerned in litigation and legal questions. The vessel will be covered by a P and I policy, as well as hull and machinery policies and pollution protection. It is therefore apparent that the underwriter is an interested party regarding a VTS operation and whether it is increasing safety in restricted waters or adding to the overall risk.

Flag, coastal and port states

Responsibility for VTS operations must also be accepted by those nations that border seaways and have an interest in shipping activity off their own coastlines.

International Maritime Organization (IMO)

IMO resolutions can only pass into law by an Act of State Legislation and it is the responsibility of IMO to ensure conformity on the part of participating nations. The COLREGS, Traffic Separation Schemes and Recommended Routeing all tend to influence the sphere of VTS but there is currently a lack of development in regulatory practice worldwide, affecting uniformity of VTS. To counter this, it is not unusual for neighbouring countries to enter into bilateral agreements to cover a VTS operation which affects both coastlines, e.g. US/Canada.

At the present time there is little case law by which a precedent can be set regarding VTS operations when they go wrong, but courts may be influenced by previous cases where navigation aids have been faulty as a result of the state's negligence. In such cases the state was held liable. However, it would seem unlikely that the Master of a vessel would be absolved of all responsibility in the event of an accident.

VTS summary

With the increased volume of shipping, and faster and larger ships becoming ever more common, the need for maritime traffic services became a natural progression to deal with shipping movements. The development of services has been considerably influenced by notable maritime accidents and related damage to the environment especially through oil pollution, and most systems now take full account of the nature of ship's cargoes.

The advantages of a VTS operation can avoid navigational conflicts by early detection and resolve developing close quarter situations before they become problematical. Risk to life and the environment can be reduced and the overall traffic flow optimized to improve throughput. Harbour movements can also be better managed to avoid congestion in confined port limits and the overall efficiency of the Port Authority can be raised with a vessel traffic service *in situ*.

In view of the detrimental effects following collision, grounding or stranding, governments have a mutual interest in the monitoring of shipping off their coastlines. They have endeavoured to provide improved navigation facilities for what is a worldwide business but the responsibilities and the liability in the case of an accident can still be somewhat vague under the state laws of the country which is controlling the system. The legal implications in many cases were considered after the VTS was established and it was only later that consequences under international law were taken into account with respect to the level of authority and regulation.

For further reading on VTS see SOLAS, Chapter V, Safety of Navigation, Regulation 8-2, Annex 20, the Dover Strait and MAREP, UK Safe Seas Guide 1 and 2, published April 1998 and British Admiralty Chart 5500 and *Vessel Traffic Systems* by Koburger.

See Figure 4.21 for a Marine Guidance Note on compliance with mandatory ship-reporting systems.

Ice seamanship

Introduction

The majority of Masters will sooner or later experience navigation in ice conditions. High latitudes present ice problems from the onset of passage planning and a Master without ice experience should be conscious of the potential hazards that would be present. Early voyage planning, with regard to the ordering of special stores, the ship's ballast arrangements, watchkeeping duties, communications, and navigational safety, are all worthy topics of consideration.

Extreme climates associated with the high latitudes and ice conditions make the working of the vessel difficult. Mechanical hatches, cargo handling gear by way of sheaves in blocks and cranes, winches and controls are all affected by freezing conditions. A general recommendation is that derricks, cranes and winches should be operated at periodic intervals to prevent freezing up if the vessel is experiencing several days of extreme cold conditions.

Ballasting in ice conditions

Unless precautions are taken by the ship's personnel regarding air pipes, sounding pipes, exposed valves, etc. freezing up and subsequent damage must be anticipated. Water tanks should have a few tonnes pumped out to ensure that no water remains in the pipes, which could more easily freeze. Warm ballast should be taken on board before the vessel enters cold climates, and once aboard should be recirculated at periodic intervals again to prevent freezing taking place.

MARINE GUIDANCE NOTE

MGN 24 (M+F)

Compliance With Mandatory Ship-reporting Systems

This note supersedes Merchant Shipping Notice M.1679

Notice to shipowners, operators, masters, officers and crew of merchant ships, skippers of fishing vessels, yachts and all other sea going craft.

1. The Merchant Shipping (Mandatory Ship-Reporting) Regulations 1996 came into force on 1 August 1996. The Regulations implement amendments to Chapter V of the Safety of Life at Sea (SOLAS) Convention allowing for the introduction of mandatory ship-reporting systems adopted by the International Maritime Organisation (IMO).

2. United Kingdom ships anywhere in the world must comply with any mandatory ship-reporting system adopted by the IMO, which applies to them.

3. The details of mandatory systems will be promulgated through the relevant parts of the Admiralty List of Radio Signals, including any amendments, corrections or replacements. Relevant entries will be annotated with the words:- '**Mandatory system under SOLAS Regulation V/8-1**'.

4. Ships to which a mandatory ship-reporting system applies should report to the shore-based authority without delay when entering and, if necessary, when leaving the area covered by the system. A ship may be required to provide additional reports or information to update or modify an earlier report.

5. Failure of a ship's radiocommunications equipment would not, in itself, be considered as a failure to comply with the rules of a mandatory ship-reporting system. However, masters should

endeavour to restore communications as soon as practicable. If a technical failure prevents a ship from reporting, the master should enter the fact and reasons for not reporting in the ship's log.

6. Masters of ships which contravene mandatory ship-reporting requirements may be liable to prosecution.

MSAS(A)
Marine Safety Agency
Spring Place
105 Commercial Road
Southampton
SO15 1EG

9 June 1997

[File ref: MNA 134/3/005]

Safe Ships Clean Seas

An executive agency of
THE DEPARTMENT
OF TRANSPORT

Figure 4.21

The Canadian Coastguard recommend some ship types de-ballast upper ballast tanks, but this may not be compatible to a Master who wants to maintain a deep draught in order to avoid rudder and propeller damage. De-ballasting also increases the windage of the vessel, and depending on current and expected weather conditions this action could expose areas of the super-structure to the increased possibility of 'ice accretion' (see Figure 4.22). The use of antifreeze fluids or salt in ballast water should also be considered as alternative options to reduce the risk of freezing. Of the two antifreeze is preferred because it is generally less corrosive than salt.

When de-ballasting it is considered better to pump out all the tank in a single operation as opposed to leaving partially full tanks. This could make the tank more susceptible to freezing and may well increase the free surface effects on the ship's stability. The stripping of residual fluids should be performed at the same time.

Ice navigation: dangers and precautions

Ice accretion

Navigating in ice regions is never easy but when sub-freezing air temperatures persist and ice accretion starts

Figure 4.22   Example of ice accretion on the deck of a general cargo vessel, adapted to take container deck cargo

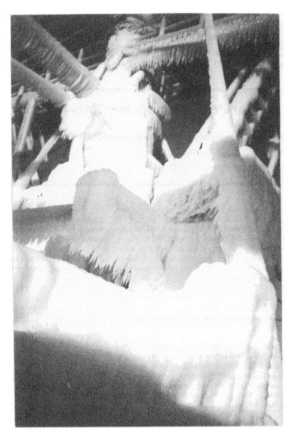

Figure 4.23   An example of ice formation that has been allowed to grow and thicken

to build on the vessel's exposed superstructure added stability problems could be generated. With severe ice accretion on the upper decks of the vessel the additional weight factor will have a direct effect on the GM of the vessel. This can be particularly pronounced with, say, a container vessel, where a 'deck container stack' could be of considerable height. It is essential that the Master attempts to reduce the level of ice accretion by altering course, if possible, to follow either a warmer climatic route, or adopt a heading that will favour a less wind chill factor. In any event a reduction in speed must be considered as a priority action while attempts are made to remove ice accumulations.

The task of breaking down and removing ice becomes a necessity on two counts: the positive stability of the vessel must be maintained and navigational instruments must be kept operational and prevented from icing up. Employing the ship's crew to break and clear ice is a cold and dangerous job. The welfare of the crew should be of concern and Masters should ensure that personnel so engaged are adequately clothed and protected against associated ice dangers. Decks will be slippery and ice

is heavy. Cracking and falling ice can lead to accidents, especially when people are cold and tired.

The use of steam hoses is an effective way of clearing ice accretion at the main deck level. However, it must be considered extremely hazardous to employ steam hoses to tackle ice accumulation in exposed high rigging positions. Adjacent formations may unexpectedly come away and crash towards deck level in areas where personnel are working. Extreme care must be taken with 'overhangs' and personnel should not be active on lower decks.

The breaking away of large ice formations not only causes ship damage but is usually difficult to handle and often awkward to break or move overside. Cracking and breaking ice formations with axes or hammers is slow work, especially where the formation has been allowed to grow and thicken (see Figure 4.23). Scuppers could well be iced up, and conditions would probably not lend to large ice blocks being left to melt and would necessitate lifting over the gunwale or side railings. Railings and freeing ports may well be blocked by ice accretion and it would be a prudent action to clear these apertures for discharge as soon as practical.

Suggested additional stores list for ice regions

- Axes and/or pick axes with spare shaft handles.
- Scrappers, manhelpers, chipping hammers, jubilee clips and bull dog grips.
- Brush heads, shovels.
- Rock salt (quantity would depend on ship's size and the ability to restock later).
- Antifreeze liquids.
- Industrial: fan heaters, electric heat lamps, and/or warm air blowers.
- Increased quantities of paraffin, gas oil, and cold start lubricants.
- Spare steam hoses with necessary couplings.
- Lagging and insulation materials.
- Canvas sheeting.
- Protective clothing, inclusive of gloves, goggles, ear muffs, insulated jackets, boiler suits, etc.
- Increased spares for engine room, boiler rooms, pumps and valves.
- Navigation and communication equipment spares, i.e. for clear view screens, heated windows.
- Extra medical stores for cold weather ailments such as colds, influenza, cold sores, and frostbite, and protective skin creams and petroleum jelly.

The quantity of equipment ordered would reflect the number of persons in the ship's complement, the size of the vessel, the expected period of duration inside the cold region and the availability and cost of goods from destination ports.

Proceeding into ice conditions

The problems for the Officer of the Watch and safe navigation of the vessel are often concerned with instruments and in particular antenna aerials and the like. Icing on a radar aerial, for example, could reduce radar energy going both outwards and inwards, resulting in poor target definition on the screen, especially as radar is considered probably the most important of all navigation instruments inside an ice region.

Coastline targets, such as headlands and islands, could be expected to appear false where ice has become 'fast' or where river mouths and estuaries have frozen over. A large iceberg could be mistaken for an island landfall or an extended headland, giving an incorrect bearing. Position monitoring should therefore always employ both primary and secondary systems when navigating inside ice regions and overreliance on one method should always be avoided.

Other instruments susceptible to icing are compass bowl face plates, which can frost over with ice build-up obscuring the compass card. Junior officers taking visual

fixes and leaving the 'helmet' off bridge wing repeaters need to learn quickly the benefits of keeping a tidy ship and putting things back in their place. The use of 'dead reckoning' is also compromised, where the log is often withdrawn when the vessel enters ice infested areas, in order to protect the instrument from damage, making a speed and distance estimate questionable.

Some on-board navigation systems may be dependent on external transmitting stations and the beacons for operation are positioned strategically to maximize reception over a given area. The fact that in winter climates and adverse weather conditions malfunctions occur more frequently, reliability of transmissions can become a problem when position fixing. Through lack of maintenance the beacon itself, set in a snow covered landscape, could effectively render the transmitter out of commission for some time.

The use of GPS, however, is unaffected in ice areas, and has become the accepted instrument alternative. Overall, the use of any position fixing method should be gainfully employed, but Masters are reminded that the use of a lookout is still considered an essential element when proceeding in ice.

Use of navigation buoys and beacons in ice conditions

Many port authorities clear channel buoys during the ice season in order to reduce damage costs, and leave a bare minimum in position. Where extreme cold temperatures exist discoloration of navigation marks, tends to occur with rusting and flaking paintwork changing the buoys' and beacons' initial appearance. This could possibly give rise to navigators misconstruing the intended function of the navigation mark. It must be remembered, however, that buoys should not be used for position fixing. If heavy ice has been present in the area floating marks could well be out of position, and the reliability of such marks must be considered suspect.

Shoreside 'sector lights' could also be expected to suffer when situated in very cold climates on or near the coastline. Ice accretion on shoreside light beacons could result in the glass/lens of the beacon being obscured. As such the effective power and associated range of the light could be drastically reduced. Any demarcation between sectors could also be confusing to a navigator when crossing a charted track with a questionable 'sector light position line'. Masters should ensure watchkeepers adhere to both a primary and secondary position fixing method and employ alternative instrument methods. See Figure 4.24.

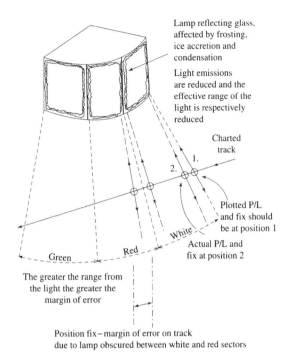

Lamp reflecting glass, affected by frosting, ice accretion and condensation

Light emissions are reduced and the effective range of the light is respectively reduced

Charted track

1.

2.

Plotted P/L and fix should be at position 1

White

Actual P/L and fix at position 2

Red

Green

The greater the range from the light the greater the margin of error

Position fix – margin of error on track due to lamp obscured between white and red sectors

Figure 4.24 Errors in the use of sector lights in ice conditions

The work of the ice breaker

Many maritime nations engage the services of ice breaker vessels in order to keep ports open and the sea lanes clear for shipping during the winter months. Over the last 60 years the coast of Finland and the Gulf of Bothnia together with other Baltic regions have experienced active developments in dual role tonnage. New ships being constructed for charter as ice breakers or offshore supply duty have eliminated the designated ice breaker lying idle for 8 months of the year.

Current tonnage has now been constructed as multi-purpose with four propellers, two forward, two aft, providing freedom of movement ahead and astern. Retractable azipod (external motor propulsion unit) thrusters, coupled with tunnel thrusters and a dynamic position management system, allow high resolution position holding when acting in the offshore arena. (Azipods as used here are without ducting nozzles, as they have been found to perform better exposed in ice conditions.)

Ice breaking features include standard ice breaker bow construction and increased scantlings together with 'rapid action rolling tanks', bubble curtains to destabilize ice formation on the hull, and a stepped hull width which is wider forward than aft, to allow the vessel to manoeuvre free if 'beset' in ice. A heli-deck landing area and remotely operated vehicle (ROV) facility provide versatility for all-year employment.

The need for ice breakers is mainly for vessels advancing in convoy (see Figure 4.25) but virtually all

Figure 4.25 Ice convoy moving through 'close pack ice' – an ice breaking lead vessel being followed by vessels in convoy at respective distance between each ship

Figure 4.26 Ice strengthened vessel proceeding slowly through 10/10ths pack ice

administrations expect vessels to be towed into clear water, should the need arise. Clear communications between participating vessels are essential and Masters should be aware that towing, if carried out, is usually conducted at short stay. This provides better control by the ice breaker. The two vessels are in close proximity, and extensive use of fenders between the towed vessel and the breaker is to be advised.

Different authorities adopt differing styles of towing, very often employing two widely spanned hawsers from either side of the towed vessel. The operation tends to generate excessive weight on these hawsers especially in heavy ice conditions, and Masters should advise their personnel of the dangers and possibility of a hawser parting unexpectedly. Ideally, ice breakers should break up the surrounding ice to allow the vessel freedom under her own accord, but this is not always possible.

See Figure 4.26 for an example of a vessel moving through 10/10ths pack ice.

5 Special maritime operations

Introduction

The role of the ship's Master is demanding enough when the vessel is operating under routine conditions. However, there are exceptional circumstances when even greater demands are required. Such occurrences could be associated with 'special operations', for example dry docking, or salvage where the need for non-routine skills becomes a requirement, or when towage becomes a necessity or where pollution of the environment is involved. These are all times which reflect the need for an experienced Master to take positive action and show the desired leadership qualities that any crisis may dictate.

The fact remains that good seamanship often means that the average Master may not experience all or any of these unusual marine incidents throughout his/her period in command, although a few Masters do have their moments. These individual experiences are not to be forgotten and are continually drawn upon when making decisions and drawing conclusions. The position of command usually requires a greater understanding of the consequences of incorrect decisions.

This chapter latterly discusses heavy lift operations and piracy, which may seem totally unrelated. Not so, they are both topics to which the Master would expect to be involved with. Routine cargo operations are one thing, a 50 or 100 tonne load affecting stability is quite another matter. Similarly, at the time of writing, piracy reflected a 40 per cent increase in incidents during 1999 over and above that in 1998, and would without doubt involve the Master directly. To be face to face with an armed pirate may happen once in a lifetime and must be considered a 'special situation'. It should certainly not be considered as routine.

Dry docking: a Master's role

When a vessel is routinely dry docked, it usually incorporates survey work and/or extensive refit work. Alternatively dry docking may take place as an essential activity because the vessel has sustained accidental damage and requires immediate repairs to remain operational in a seaworthy condition. The role of the Master is to bring the vessel safely onto the blocks and ensure that the owner's interests throughout the dry dock period are protected. If the vessel is listed because of hull damage with possible loss of watertight integrity, the ship handling skills and general seamanship required would make the delivery task to the blocks more demanding.

It is not unusual for the ship's Chief Officer to calculate the vessel's stability but the prudent Master should be keen to monitor that the ship has an adequate *positive GM* and is at a suitable trim to comply with the Dry Dock Authority. In routine docking with the bow first, the vessel is upright and trimmed slightly by the stern. When docking with a vessel in a damaged state, these conditions cannot always be guaranteed and she may be required to enter the dock stern first, dependent on the location of any damage.

Alignment of the vessel should be achieved in conjunction with the co-operation of the Dry Dock Manager who may employ docking 'plumb bobs' to line up the keel on the centre blocks, and may additionally employ divers to confirm the desired position. Use of moorings to port and starboard, to hold station, is expected once alignment is established, prior to pumping operations being commenced.

While the vessel remains afloat the stability will be unimpaired, but once the ship physically touches the blocks the positive stability will start to deteriorate and become progressively worse as the water line falls. A destabilizing effect is generated by the upthrust from the blocks and this upthrust is proportional to the lost buoyancy. As the water level continues to fall the vessel's trim will reduce until she becomes 'sewn' overall on the blocks, and at this moment stability no longer remains a problem.

Definitions

'Critical instant (dry docking)' is defined as that moment when the ship's hull makes contact with the blocks and the stability of the vessel is at its worst condition prior to landing overall.

'Critical period (dry docking)' is defined as that interval of time between the vessel first touching the blocks and landing overall (the fully landed vessel is described as being 'sewn' on the blocks).

Dry dock period

The list of maintenance activities for a ship in dry dock can be endless and it is debatable whether the Master would be directly involved. When a good Chief Officer is employed the load on the Master can theoretically be somewhat relieved. However, that same industrious First Mate with or without limited dry dock experience could be reassured by the presence of the Master in an advisory role.

It is also the only time seagoing personnel have the opportunity to examine, first hand, the underwater area of the hull and the ship's Master would be unwise to miss the experience. It is a time that allows detailed inspection of the rudder(s), the keel, bottom plating, stabilizing fins, bow thrust units, etc., and the period is one which Masters should encourage junior personnel to observe to learn more about their chosen profession (see Figure 5.1).

Utility requirements

In order to maintain continuity regarding safety procedures the Master or Chief Officer should ensure that the following utilities are available for the ship as soon after arrival in the dock as possible. Although the Master may not actually request the various services he would most certainly be instrumental in ensuring the company or the Chief Officer obtain necessary facilities for the well-being of the vessel and all on board.

1. *Water* – for pressurizing the ship's fire main. It may be a requirement that the 'international shore connection' is employed to effect this.
2. *Access* – one or two gangways, supplied usually by the Dry Dock Authority. Although supplied by a shoreside authority, the responsibility for providing a safe means of access to the shore remains clearly in the domain of the Master. It is the Master's duty to instruct ship's personnel, and in particular the duty officer, to monitor the gangway condition throughout the dry dock period.

   NB. In the event of a fire on board the vessel, in close proximity to the gangway, if no alternative (second gangway) is available, then personnel might not be able to evacuate the vessel if required.

Figure 5.1   A dry docked vessel, allowing ship's personnel, among others, the opportunity to observe and inspect rudder(s), keel, bottom plating, etc.

3. *Garbage* – a shoreside 'skip' should be made available for general garbage as soon as practical. Garbage retained on board becomes an immediate health hazard and tends to encourage vermin.
4. *Communications* – many dry docks are often located well clear of local communities and consequently effective communications can be poor. The need to bring in the fire brigade or ambulance service may become essential and poor communications could have fatal consequences. Masters must ensure that the ship is equipped with an effective communication link for emergency use, especially important if use of the ship's radio is curtailed when inside harbour limits.
5. *Security* – dry dock basins are notorious for breaches in security and what would normally be considered good housekeeping or, in this case, good shipkeeping practice. Gangway security, with a regular deck/fire patrol operation, is recognized as an obvious deterrent against law breakers and accidents occurring on board the ship, e.g. nightwatchman service.

6. *Sanitation* – toilet facilities are a necessity of life. If the crew are continuing to live on board, as opposed to being billeted in hotel accommodation ashore, basic sanitation needs must be catered for. With this in mind the ship may have the ability to use its own fresh water supply, but alternatively 'domestic water' would need to be laid on by the Dry Dock Authority, as a basic requirement. Masters need to satisfy themselves that basic needs are being made available.

NB. Modern tonnage may have the use of self-contained internal sewage systems but in the absence of such a system, shoreside toilets are usually a standard feature of most dry docks and are made available for ships' crews.

7. *Power* – an electrical power supply for the operation of general services, heating, lighting, monitoring instrumentation, etc. is clearly an essential need and cannot always be delivered from shipboard equipment when engine rooms are closed down for even a short period. Dry docks are aware that the need for power, if only to keep the workforce active, must be continuously available to the ship and portable (shoreside) generators are usually laid on to provide adequate power for the ship's needs. The prudent Master should note that when generators are supplied by the authorities, they should normally also couple up a 'bonding wire' to reduce the risk of a static electrical build-up.

NB. Additionally, a pneumatic line may be needed, depending on the the needs and the usage requirements of the vessel.

Figure 5.3 Removal of damaged propeller from tail end shaft

Figure 5.4 *C.S. Nexus,* a cable laying vessel on charter with Cable & Wireless, lays in a graving dry dock. Midships gangway access can be seen on the foredeck. Overside paintwork is completed, while maintenance is continuing on bow thrust units. Bilge keel is visible from the midships length

Figures 5.2, 5.3 and 5.4 show a range of ongoing, dry dock operations.

Dry dock crew management

With the many maintenance tasks required aboard a vessel in dry dock Masters or the officer in charge must be concerned with exceptional circumstances which could affect normal safety procedures. Extensive burning, electrical work, or enclosed space entry are considered the norm when in dock and all such activity must be covered by a 'permit to work' system.

A briefing of officers and crew regarding safety awareness and the dangers of not complying with recommended practice should be delivered and understood by all personnel, bearing in mind that over any dry dock

Figure 5.2 Dry dock showing a line of keel blocks. Ongoing operations include ranging of cables, cable remarking and hull and bulbous bow painting

period many contractors are employed to carry out specific functions and may not possess the same degree of safety training inherent with seafarers. Their function could overlap with crew activity and adversely influence safe practice aboard the vessel. (Risk assessment required.)

Dry dock departure

It is essential that prior to flooding the dry dock, the ship's stability is recalculated. Maintenance activities while in dock very often involve the deballasting of tanks, or the cleaning of fresh water tanks, commodity tanks, or specialized compartments. Any movement of 'weight' on board will affect the *on the block soundings* and it is imperative that the vessel is returned to her original condition of entry when afloat or similar upright condition prior to flooding of the dock.

NB. A real danger of the vessel not being in equilibrium exists if the ship's tanks are not set in a suitable condition of loading so as to prevent the vessel slipping from the blocks during the refloating period.

Authority to Flood Certificate

The Master or his representative, at some time near to completion of the dry dock period, will be asked by the Dry Dock Manager to sign the Authority to Flood Certificate. It would be foolhardy to sign such a document without first thoroughly checking that all survey work has been completed, and it is especially important to ensure oneself that any tank plugs drawn have been replaced.

A walk around the bottom of the dock, to sight plugs in position, can be reassuring. At the same time when leaving the dry dock check that no persons are remaining in what will soon become a flooded area, prior to signing this certificate.

Floating dry docks

The fundamental difference between a land-based dock and a floating dry dock is that the floating dock is a tanking system which can be submerged to allow the vessel's entry. The tank system of the floating dock allows the dock itself to be trimmed or listed to accommodate the ship, especially desirable if the vessel is damaged or listing.

Salvage activities

Marine casualties continue to occur worldwide, and salvage services have generally become a commercial business following such incidents. The majority of the salvage companies are members of the International Salvage Union (ISU) and are located in many countries worldwide.

Salvage cases in British waters are normally conducted by the British Admiralty Courts and are subject to the International Salvage Convention of 1989 which has been part of English statute law since 1 January 1995, under the provision of the Merchant Shipping (Salvage and Pollution) Act 1994.

The principle of salvage

When marine property is recognized as being in danger from whatever source, voluntary or contractual services may be rendered to safeguard and preserve that property, and, if successful, persons involved in recovery operations may claim a reward based on the clause – 'no cure, no pay' in other words, property must be preserved in order for a reward to be payed out.

The Master's role and salvage

Under the International Salvage Convention 1989 the Master has the authority to negotiate and complete a contract on behalf of the shipowner. The Master and/or the shipowner similarly have the authority to complete a contract on behalf of the owner of any goods on board the vessel. These provisions, which could become relevant in the event of an emergency, are important changes. Previous clauses allowed the shipowners to insist that the Master did not accept salvage services and the convention now prevents this happening, as a matter of law.

Clearly the first to be aware of a marine casualty are the Master and crew of the affected vessel (or officer in charge in the event of the death of the Master). Sophisticated communications remove any ambiguity regarding decision making and consultation has become the order of the day. Masters would therefore duly contact owners, insurers, etc., while being fully aware that the safety of the ship, property on board and the welfare of crew and/or passengers would rest solely with himself as the ship's Master.

Salvage operations: documentation

Once a Master or shipowner realizes that their ship is in trouble they are expected to pay the financial costs to safeguard the vessel. When salvage claims are involved it is essential that the owner's interests are protected and Masters should use any or all of the following to ensure that a true account of events will be presented to the courts or to arbitrators: video evidence, tape recordings, photographs, witnessed statements and the contemporaneous records of the vessel such as charts, log books, movement books, etc.

Figure 5.5 The *Herald of Free Enterprise* lying secured alongside after a successful salvage operation which involved 'parbuckling' to right the capsized vessel off Zeebrugge (1987)

Salvage services and the subsequent remuneration will be determined by the contract and in accordance with the law or by terms specified by Lloyd's Open Form (LOF), see Figure 5.6, initially introduced in 1890. At the time of writing LOF '95 is applicable and describes the duties of the parties concerned. Agreement can be achieved by radio between those involved and subsequent administrative forms can be signed at a later time.

NB. LOF '95 carries the amendments as specified by the salvage convention and subsequently does not bring the agreement into conflict with English law.

Types of salvage

1. *Conventional Salvage.* This is probably the most widely used by a ship's Master and covers the services provided when a vessel has become a casualty through grounding, fire, or being immobilized and unable to proceed under its own power. The principle being to return the vessel to regular service following necessary repairs.
2. *Port clearance or wreck removal.* This is generally a service which is carried out to benefit safe navigation and usually conducted inside port limits or around oilfields or environmentally protected areas. Generally, it is ordered by a statutory body like a government agency or the Port Authority which can order the removal of debris, a vessel or its cargo.

NB. Redundant oil/gas installations fall into this category.

3. *Salvage of cargo.* Cargo salvage occurs usually when the hull is so badly damaged that the ship itself cannot be economically salved. However, the cargo or part of the cargo can be recovered by a salvage operation.
4. *Underwater salvage (wet salvage).* The costs incurred with recovery of the vessel and/or its cargo are usually very high, so much so that the recovery of the vessel itself is often not a viable proposition. Depending on the type of cargo and the method of recovery it may be possible to recover the cargo economically but not the vessel.
5. *Archaeological salvage.* This has become probably one of the most popular types of salvage operations by persons with both legal and unfortunately illegal intentions. The recovery of artefacts has caused governments to act in ways to preserve wreck sites or become directly involved themselves, under contract, to recover for prosperity. The growth of this activity has been coupled with an increased use of scuba-diving and as with any underwater activity many underestimate the associated dangers, particularly relevant when amateur divers are engaged below the surface and unaware of on-surface activity.

Methods of vessel recovery

1. When righting and refloating a capsized vessel within the confines of a dock, the number of cantilever frames secured to the external hull and the number of anchored winches fixed into the dock side depends on the vessel's length and weight.
2. Recovery of a smaller type vessel can be achieved by employing the principle of 'parbuckling'. This is achieved by the use of a floating crane/sheer leg with wire strops around the vessel to cause rotation and lift.

NB. Care must be exercised not to turn the vessel onto its other side.

3. Large vessel recovery, e.g. *Herald of Free Enterprise*, in open water conditions can be achieved by using sheer leg floating barges with pulling barges. Anchor points for pulling barges are provided by piles into the seabed. Reaction anchors stabilize rotation.

Salvage operations

A great variety of operations can be covered by a salvage agreement, and many involve the removal of cargo before the ship itself can be salved. It would not be unrealistic for the value of the cargo to exceed the value of the vessel and each operation would be determined on

LOF 1995

LLOYD'S

STANDARD FORM OF

SALVAGE AGREEMENT

(APPROVED AND PUBLISHED BY THE COUNCIL OF LLOYD'S)

NO CURE - NO PAY

On board the.................................

Dated.....................

IT IS HEREBY AGREED between Captain......................................

for and on behalf of the Owners of the "....................................." her
cargo freight bunkers stores and any other property thereon (hereinafter collectively called "the Owners")
and...for and on behalf of..
.....................................(hereinafter called "the Contractor"*) that:-

1. (a) The Contractor shall use his best endeavours:-

(i) to salve the "....................................."and/or her cargo freight bunkers
stores and any other property thereon and take them to #...................................or
to such other place as may hereafter be agreed either place to be deemed a place of safety or if no such
place is named or agreed to a place of safety and
(ii) while performing the salvage services to prevent or minimize damage to the environment.

(b) Subject to the statutory provisions relating to special compensation the services shall be rendered and
accepted as salvage services upon the principle of "no cure - no pay."

(c) The Contractor's remuneration shall be fixed by Arbitration in London in the manner hereinafter
prescribed and any other difference arising out of this Agreement or the operations thereunder shall be
referred to Arbitration in the same way.

(d) In the event of the services referred to in this Agreement or any part of such services having been
already rendered at the date of this Agreement by the Contractor to the said vessel and/or her cargo
freight bunkers stores and any other property thereon the provisions of this Agreement shall apply to
such services.

(e) The security to be provided to the Council of Lloyd's (hereinafter called "the Council") the Salved
Value(s) the Award and/or any Interim Award(s) and/or any Award on Appeal shall be in
#... currency.

(f) If clause 1(e) is not completed then the security to be provided and the Salved Value(s) the Award
and/or Interim Award(s) and/or Award on Appeal shall be in Pounds Sterling.

(g) This Agreement and Arbitration thereunder shall except as otherwise expressly provided be governed
by the law of England, including the English law of salvage.

PROVISIONS AS TO THE SERVICES

2. *Definitions:* In this Agreement any reference to "Convention" is a reference to the International Convention
on Salvage 1989 as incorporated in the Merchant Shipping (Salvage and Pollution) Act 1994 (and any
amendment thereto). The terms "Contractor" and "services"/"salvage services" in this Agreement shall have the
same meanings as the terms "salvor(s)" and "salvage operation(s)" in the Convention.

3. *Owners Cooperation:* The Owners their Servants and Agents shall co-operate fully with the Contractor in
and about the salvage including obtaining entry to the place named or the place of safety as defined in clause 1.
The Contractor may make reasonable use of the vessel's machinery gear equipment anchors chains stores and
other appurtenances during and for the purpose of the salvage services free of expense but shall not
unnecessarily damage abandon or sacrifice the same or any property the subject of this Agreement.

4. *Vessel Owners Right to Terminate:* When there is no longer any reasonable prospect of a useful result
leading to a salvage reward in accordance with Convention Article 13 the owners of the vessel shall be entitled
to terminate the services of the Contractor by giving reasonable notice to the Contractor in writing.

PROVISIONS AS TO SECURITY

5. (a) The Contractor shall immediately after the termination of the services or sooner notify the Council
and where practicable the Owners of the amount for which he demands salvage security (inclusive of costs
expenses and interest) from each of the respective Owners.

(b) Where a claim is made or may be made for special compensation, the owners of the vessel shall on
the demand of the Contractor whenever made provide security for the Contractor's claim for special
compensation provided always that such demand is made within two years of the date of termination of the
services.

(c) The amount of any such security shall be reasonable in the light of the knowledge available to the
Contractor at the time when the demand is made. Unless otherwise agreed such security shall be provided (i) to
the Council (ii) in a form approved by the Council and (iii) by persons firms or corporations either acceptable to
the Contractor or resident in the United Kingdom and acceptable to the Council. The Council shall not be
responsible for the sufficiency (whether in amount or otherwise) of any security which shall be provided nor the
default or insolvency of any person firm or corporation providing the same.

(d) The owners of the vessel their Servants and Agents shall use their best endeavours to ensure that the
cargo owners provide their proportion of salvage security before the cargo is released.

6. (a) Until security has been provided as aforesaid the Contractor shall have a maritime lien on the
property salved for his remuneration.

(b) The property salved shall not without the consent in writing of the Contractor (which shall not be
unreasonably withheld) be removed from the place to which it has been taken by the Contractor under clause
1(a). Where such consent is given by the Contractor on condition that the Contractor is provided with
temporary security pending completion of the voyage the Contractor's maritime lien on the property salved shall
remain in force to the extent necessary to enable the Contractor to compel the provision of security in
accordance with clause 5(c).

(c) The Contractor shall not arrest or detain the property salved unless:-

(i) security is not provided within 14 days (exclusive of Saturdays and Sundays or other days
observed as general holidays at Lloyd's) after the date of the termination of the services or
(ii) he has reason to believe that the removal of the property salved is contemplated contrary to
clause 6(b) or
(iii) any attempt is made to remove the property salved contrary to clause 6(b).

(d) The Arbitrator appointed under clause 7 or the Appeal Arbitrator(s) appointed under clause 13(d)
shall have power in their absolute discretion to include in the amount awarded to the Contractor the whole or
part of any expenses reasonably incurred by the Contractor in:-

(i) ascertaining demanding and obtaining the amount of security reasonably required in accordance
with clause 5.
(ii) enforcing and/or protecting by insurance or otherwise or taking reasonable steps to enforce
and/or protect his lien.

Figure 5.6

PROVISIONS AS TO ARBITRATION

7. (a) Whether security has been provided or not the Council shall appoint an Arbitrator upon receipt of a written request made by letter telex facsimile or in any other permanent form provided that any party requesting such appointment shall if required by the Council undertake to pay the reasonable fees and expenses of the Council and/or any Arbitrator or Appeal Arbitrator(s).

(b) Where an Arbitrator has been appointed and the parties do not proceed to arbitration the Council may recover any fees costs and/or expenses which are outstanding.

8. The Contractor's remuneration and/or special compensation shall be fixed by the Arbitrator appointed under clause 7. Such remuneration shall not be diminished by reason of the exception to the principle of "no cure - no pay" in the form of special compensation.

REPRESENTATION

9. Any party to this Agreement who wishes to be heard or to adduce evidence shall nominate a person in the United Kingdom to represent him failing which the Arbitrator or Appeal Arbitrator(s) may proceed as if such party had renounced his right to be heard or adduce evidence.

CONDUCT OF THE ARBITRATION

10. (a) The Arbitrator shall have power to:-

(i) admit such oral or documentary evidence or information as he may think fit

(ii) conduct the Arbitration in such manner in all respects as he may think fit subject to such procedural rules as the Council may approve

(iii) order the Contractor in his absolute discretion to pay the whole or part of the expense of providing excessive security or security which has been unreasonably demanded under Clause 5(b) and to deduct such sum from the remuneration and/or special compensation

(iv) make Interim Award(s) including payment(s) on account on such terms as may be fair and just

(v) make such orders as to costs fees and expenses including those of the Council charged under clauses 10(b) and 14(b) as may be fair and just.

(b) The Arbitrator and the Council may charge reasonable fees and expenses for their services whether the Arbitration proceeds to a hearing or not and all such fees and expenses shall be treated as part of the costs of the Arbitration.

(c) Any Award shall (subject to Appeal as provided in this Agreement) be final and binding on all the parties concerned whether they were represented at the Arbitration or not.

INTEREST & RATES OF EXCHANGE

11. *Interest:* Interest at rates per annum to be fixed by the Arbitrator shall (subject to Appeal as provided in this Agreement) be payable on any sum awarded taking into account any sums already paid:-

(i) from the date of termination of the services unless the Arbitrator shall in his absolute discretion otherwise decide until the date of publication by the Council of the Award and/or Interim Award(s) and

(ii) from the expiration of 21 days (exclusive of Saturdays and Sundays or other days observed as general holidays at Lloyd's) after the date of publication by the Council of the Award and/or Interim Award(s) until the date payment is received by the Contractor or the Council both dates inclusive.

For the purpose of sub-clause (ii) the expression "sum awarded" shall include the fees and expenses referred to in clause 10(b).

12. *Currency Correction:* In considering what sums of money have been expended by the Contractor in rendering the services and/or in fixing the amount of the Award and/or Interim Award(s) and/or Award on Appeal the Arbitrator or Appeal Arbitrator(s) shall to such an extent and in so far as it may be fair and just in all the circumstances give effect to the consequences of any change or changes in the relevant rates of exchange which may have occurred between the date of termination of the services and the date on which the Award and/or Interim Award(s) and/or Award on Appeal is made.

PROVISIONS AS TO APPEAL

13. (a) Notice of Appeal if any shall be given to the Council within 14 days (exclusive of Saturdays and Sundays or other days observed as general holidays at Lloyd's) after the date of the publication by the Council of the Award and/or Interim Award(s).

(b) Notice of Cross-Appeal if any shall be given to the Council within 14 days (exclusive of Saturdays and Sundays or other days observed as general holidays at Lloyd's) after notification by the Council to the parties of any Notice of Appeal. Such notification if sent by post shall be deemed received on the working day following the day of posting.

(c) Notice of Appeal or Cross-Appeal shall be given to the Council by letter telex facsimile or in any other permanent form.

(d) Upon receipt of Notice of Appeal the Council shall refer the Appeal to the hearing and determination of the Appeal Arbitrator(s) selected by it.

(e) If any Notice of Appeal or Cross-Appeal is withdrawn the Appeal hearing shall nevertheless proceed in respect of such Notice of Appeal or Cross-Appeal as may remain.

(f) Any Award on Appeal shall be final and binding on all the parties to that Appeal Arbitration whether they were represented either at the Arbitration or at the Appeal Arbitration or not.

CONDUCT OF THE APPEAL

14. (a) The Appeal Arbitrator(s) in addition to the powers of the Arbitrator under clauses 10(a) and 11 shall have power to:-

(i) admit the evidence which was before the Arbitrator together with the Arbitrator's notes and reasons for his Award and/or Interim Award(s) and any transcript of evidence and such additional evidence as he or they may think fit.

(ii) confirm increase or reduce the sum awarded by the Arbitrator and to make such order as to the payment of interest on such sum as he or they may think fit.

(iii) confirm revoke or vary any order and/or Declaratory Award made by the Arbitrator.

(iv) award interest on any fees and expenses charged under paragraph (b) of this clause from the expiration of 21 days (exclusive of Saturdays and Sundays or other days observed as general holidays at Lloyd's) after the date of publication by the Council of the Award on Appeal and/or Interim Award(s) on Appeal until the date payment is received by the Council both dates inclusive.

(b) The Appeal Arbitrator(s) and the Council may charge reasonable fees and expenses for their services in connection with the Appeal Arbitration whether it proceeds to a hearing or not and all such fees and expenses shall be treated as part of the costs of the Appeal Arbitration.

PROVISIONS AS TO PAYMENT

15. (a) In case of Arbitration if no Notice of Appeal be received by the Council in accordance with clause 13(a) the Council shall call upon the party or parties concerned to pay the amount awarded and in the event of non-payment shall subject to the Contractor first providing to the Council a satisfactory Undertaking to pay all the costs thereof realize or enforce the security and pay therefrom to the Contractor (whose receipt shall be a good discharge to it) the amount awarded to him together with interest if any. The Contractor shall reimburse the parties concerned to such extent as the Award is less than any sums paid on account or in respect of Interim Award(s).

(b) If Notice of Appeal be received by the Council in accordance with clause 13 it shall as soon as the Award on Appeal has been published by it call upon the party or parties concerned to pay the amount awarded and in the event of non-payment shall subject to the Contractor first providing to the Council a satisfactory Undertaking to pay all the costs thereof realize or enforce the security and pay therefrom to the Contractor (whose receipt shall be a good discharge to it) the amount awarded to him together with interest if any. The Contractor shall reimburse the parties concerned to such extent as the Award on Appeal is less than any sums paid on account or in respect of the Award or Interim Award(s).

(c) If any sum shall become payable to the Contractor as remuneration for his services and/or interest and/or costs as the result of an agreement made between the Contractor and the Owners or any of them the Council in the event of non-payment shall subject to the Contractor first providing to the Council a satisfactory Undertaking to pay all the costs thereof realize or enforce the security and pay therefrom to the Contractor (whose receipt shall be a good discharge to it) the said sum.

(d) If the Award and/or Interim Award(s) and/or Award on Appeal provides or provides that the costs of the Arbitration and/or of the Appeal Arbitration or any part of such costs shall be borne by the Contractor such costs may be deducted from the amount awarded or agreed before payment is made to the Contractor unless satisfactory security is provided by the Contractor for the payment of such costs.

Figure 5.6 *(continued)*

(e) Without prejudice to the provisions of clause 5(c) the liability of the Council shall be limited in any event to the amount of security provided to it.

GENERAL PROVISIONS

16. *Scope of Authority*: The Master or other person signing this Agreement on behalf of the property to be salved enters into this Agreement as agent for the vessel her cargo freight bunkers stores and any other property thereon and the respective Owners thereof and binds each (but not the one for the other or himself personally) to the due performance thereof.

17. *Notices*: Any Award notice authority order or other document signed by the Chairman of Lloyd's or any person authorised by the Council for the purpose shall be deemed to have been duly made or given by the Council and shall have the same force and effect in all respects as if it had been signed by every member of the Council.

18. *Sub-Contractor(s)*: The Contractor may claim salvage and enforce any Award or agreement made between the Contractor and the Owners against security provided under clause 5 or otherwise if any on behalf of any Sub-Contractors his or their Servants or Agents including Masters and members of the crews of vessels employed by him or by any Sub-Contractors in the services provided that the first provides a reasonably satisfactory indemnity to the Owners against all claims by or liabilities to the said persons.

19. *Inducements prohibited*: No person signing this Agreement or any party on whose behalf it is signed shall at any time or in any manner whatsoever offer provide make give or promise to provide demand or take any form of inducement for entering into this Agreement.

For and on behalf of the Contractor	For and on behalf of the Owners of property to be salved.
(To be signed by the Contractor personally or by the Master of the salving vessel or other person whose name is inserted in line 4 of this Agreement)	(To be signed by the Master or other person whose name is inserted in line 1 of this Agreement)

INTERNATIONAL CONVENTION ON SALVAGE 1989

The following provisions of the Convention are set out below for information only.

Article 1

Definitions

(a) *Salvage operation* means any act or activity undertaken to assist a vessel or any other property in danger in navigable waters or in any other waters whatsoever

(b) *Vessel* means any ship or craft, or any structure capable of navigation

(c) *Property* means any property not permanently and intentionally attached to the shoreline and includes freight at risk

(d) *Damage to the environment* means substantial physical damage to human health or to marine life or resources in coastal or inland waters or areas adjacent thereto, caused by pollution, contamination, fire, explosion or similar major incidents

(e) *Payment* means any reward, remuneration or compensation due under this Convention

Article 6

Salvage Contracts

1. This Convention shall apply to any salvage operations save to the extent that a contract otherwise provides expressly or by implication

2. The master shall have the authority to conclude contracts for salvage operations on behalf of the owner of the vessel. The master or the owner of the vessel shall have the authority to conclude such contracts on behalf of the owner of the property on board the vessel

Article 8

Duties of the Salvor and of the Owner and Master

1. The salvor shall owe a duty to the owner of the vessel or other property in danger:

(a) to carry out the salvage operations with due care;

(b) in performing the duty specified in subparagraph (a), to exercise due care to prevent or minimize damage to the environment;

(c) whenever circumstances reasonably require, to seek assistance from other salvors; and

(d) to accept the intervention of other salvors when reasonably requested to do so by the owner or master of the vessel or other property in danger; provided however that the amount of his reward shall not be prejudiced should it be found that such a request was unreasonable

2. The owner and master of the vessel or the owner of other property in danger shall owe a duty to the salvor:

(a) to co-operate fully with him during the course of the salvage operations;

(b) in so doing, to exercise due care to prevent or minimize damage to the environment; and

(c) when the vessel or other property has been brought to a place of safety, to accept redelivery when reasonably requested by the salvor to do so

Article 13

Criteria for fixing the reward

1. The reward shall be fixed with a view to encouraging salvage operations, taking into account the following criteria without regard to the order in which they are presented below:

(a) the salved value of the vessel and other property;

(b) the skill and efforts of the salvors in preventing or minimizing damage to the environment;

(c) the measure of success obtained by the salvor;

(d) the nature and degree of the danger;

(e) the skill and efforts of the salvors in salving the vessel, other property and life;

(f) the time used and expenses and losses incurred by the salvors;

(g) the risk of liability and other risks run by the salvors or their equipment;

(h) the promptness of the services rendered;

(i) the availability and use of vessels or other equipment intended for salvage operations;

(j) the state of readiness and efficiency of salvor's equipment and the value thereof

2. Payment of a reward fixed according to paragraph 1 shall be made by all of the vessel and other property interests in proportion to their respective salved values

3. The rewards, exclusive of any interest and recoverable legal costs that may be payable thereon, shall not exceed the salved value of the vessel and other property

Article 14

Special Compensation

1. If the salvor has carried out salvage operations in respect of a vessel which by itself or its cargo threatened damage to the environment and has failed to earn a reward under Article 13 at least equivalent to the special compensation assessable in accordance with this Article, he shall be entitled to special compensation from the owner of that vessel equivalent to his expenses as herein defined

2. If, in the circumstances set out in paragraph 1, the salvor by his salvage operations has prevented or minimized damage to the environment, the special compensation payable by the owner to the salvor under paragraph 1 may be increased up to a maximum of 30% of the expenses incurred by the salvor. However, the Tribunal, if it deems it fair and just to do so and bearing in mind the relevant criteria set out in Article 13, paragraph 1, may increase such special compensation further, but in no event shall the total increase be more than 100% of the expenses incurred by the salvor

3. Salvor's expenses for the purpose of paragraphs 1 and 2 means the out-of-pocket expenses reasonably incurred by the salvor in the salvage operation and a fair rate for equipment and personnel actually and reasonably used in the salvage operation, taking into consideration the criteria set out in Article 13, paragraph 1(h), (i) and (j)

4. The total special compensation under this Article shall be paid only if and to the extent that such compensation is greater than any reward recoverable by the salvor under Article 13

5. If the salvor has been negligent and has thereby failed to prevent or minimize damage to the environment, he may be deprived of the whole or part of any special compensation due under this Article

6. Nothing in this Article shall affect any right of recourse on the part of the owner of the vessel

Figure 5.6 (*continued*)

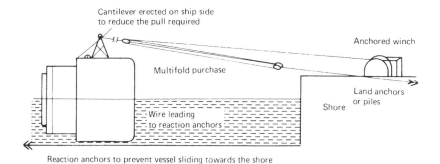

Figure 5.7   Righting and refloating a capsized vessel within the confines of a dock

its merits, usually by the three parties directly concerned, namely:

1. The shipowner who would assess the cost of salvage, repair and associated fees as to whether they would exceed the value of the vessel. Any interested third parties, such as cargo owners, charterers, mortgagees, etc., would also have to be consulted regarding their interest in the vessel.
2. The salvor who would consider the costs of the operation prior to offering salvage services on any salvage agreement or on a 'no cure, no pay' basis. Any costs incurred by the salvor would be weighed against the eventual gain of the property salved and whether such an operation would render an eventual profit.
3. The underwriters or their representatives who would assess the cost of salvage and repairs and consider whether these would exceed the insured value stipulated in the insurance policy prior to agreement to enter into a salvage contract.

Ideally if the vessel could be salved with the cargo on board this is usually the preferred option. Cargo handling and transshipment could be reduced and overall expenses would be considerably less than when separating the cargo from the vessel. Clearly if the vessel is afloat and in a reasonable condition following the casualty, it may be towed to a destination where the value of cargo and vessel can be correctly addressed.

Under previous Lloyd's Open Form agreements the responsibility to exercise due care to prevent or minimize damage to the environment was placed under the broader duty of the salvage operator, and the salvor's skill and efforts would then be taken into account when apportioning reward. Following the International Salvage Convention of 1989, LOF '95, which incorporates the Merchant Shipping (Salvage and pollution) Act 1994, became effective in January 1995. This is basically an Article titled 'Special Compensation' which may be allotted to the salvor, amounting to 30 per cent of his

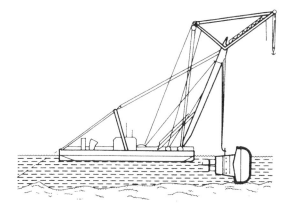

Figure 5.8   'Parbuckling'

expenses if he is successful in protecting the environment. In certain circumstances this compensation may be increased, but would not exceed 100 per cent of the salvor's expenses.

NB. Many casualties unfortunately involve oil pollution and it is in the public interest to protect the marine environment. Following the grounding of the vessel '*Braer*' in 1993 off the Shetland Islands a report by Lord Donaldson ('Safer Ships, Cleaner Seas') recommended the government retain strategic salvage cover around the UK coastline.

Removal of cargo

Ship to ship transfer

This method can be employed for a variety of cargoes including bulk commodities, liquid cargoes, containers or general cargo. Geography, especially depth of water around a casualty, will usually dictate how close a lightening vessel can approach to effect cargo transfer. Ideally if support craft can come alongside the stricken vessel then lifting gear from either vessel may be brought into use, assuming that operational power is

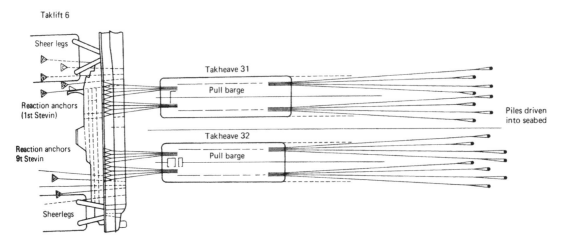

Figure 5.9   Large vessel recovery

available to engage the lifting gear aboard the stricken vessel. Many bulk carriers do not carry lifting gear and if cargo transfer is to be effected then the lightening vessel would need to be positioned so as to engage its own cranes/grabs, etc. (see Figure 5.10).

Ship to ship transfer is a particular salvage operation where ship's Masters could expect to be on scene and directly involved. Transshipment costs are high and stability calculations become essential when part cargoes are discharged, for example when a vessel is to attempt refloating after a grounding incident. Also the type of cargo transfer could well have safety implications for the well-being of the crew and other personnel on board.

Ship to ship transfer: liquid cargoes

The use of shuttle tankers to reduce pollution and lighten large tanker vessels has become a regular activity following marine casualties. If the lightening vessel can draw alongside then pumping can be carried out via the manifolds of the two vessels without too many problems. If the incident is a non-tanker vessel, and bunkers are to be transferred, the use of portable pumps or systems within a vessel's engine space are very often utilized. Where a vessel is aground concern must be addressed to the fact that as the weight transfer takes place, the draught of the stricken vessel could change and the vessel could start to refloat itself unintentionally. This in effect could have serious additional pollution implications and could effectively move the casualty from the aground position into deeper water while still in a damaged state. To prevent this, heavy duty 'ground tackle' is generally laid before lightening of the casualty is allowed to commence.

Where an approach to the casualty is restricted, probably by lack of underkeel clearance, flexible oil bearing pipelines can be employed to effect liquid transfer. In

such cases 'ground tackle' is often used for both the casualty vessel and the lightening shuttle vessel, in order to steady movement on the flexible oil bearing pipelines.

Potential hazards

Whenever, chemicals, oils or gases are being transferred there are inherent problems due to the possible levels of exposure to gases and/or toxics on deck, especially if the hull of the carrier is breached. Professional salvors tend to be fully equipped and familiar with breathing apparatus, gas detection instruments and the like. Portable pumping equipment, if employed, generally takes into account the need to be safe in volatile atmospheres and safe working practices become a necessity. Fire risks are always present and in such circumstances tend to be increased considerably.

Where floating, oil bearing pipelines are used, it is essential that continuous weather monitoring takes place. Weather conditions play an important part in any salvage operation, and become particularly relevant when there is a risk to the environment from damaged oil lines.

Where a ship is being considered a total loss, but salvage is being attempted on the cargo, it is not unusual to find that much of the equipment from the vessel itself has been removed. Crews and salvage personnel alike will need to consider the available resources remaining, in the event of emergency, preferably before such an emergency arises. Being prepared with items of equipment such as oil booms, dispersal chemicals and/or skimmer vessels is essential for operations involving oil cargoes.

Use of flexible oil bearing (floating) lines

Figure 5.11 shows a lightening vessel, restricted by available depth of water and prevented from drawing alongside the grounded vessel. A stern mooring is stretched between the grounded vessel and the relieving

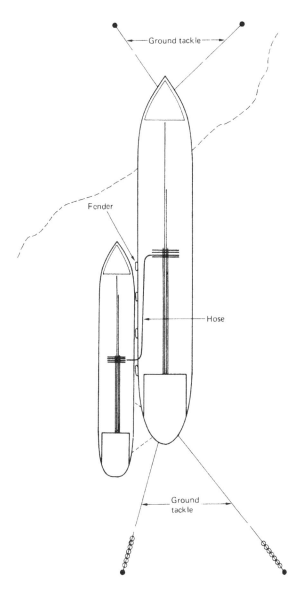

Figure 5.10 Shuttle (lightening) tanker alongside a larger vessel which has grounded. This operation is usually carried out with the view to refloating the grounded vessel at a suitable tide

shuttle tanker as well as ground tackle laid to restrict movement of the lightening vessel during the operation. The forward end of the lightening vessel employs its own anchors to hold off the shoal and further reduce movement when the draught starts to change on the vessel.

Salvage of cargo

The salvage of cargo is effected where either the vessel is not economically recoverable or where the cargo must first be removed to allow the vessel to be refloated

(sometimes referred to as the 'smash and grab' method, see Figure 5.12).

There are other methods of salving cargo.

Lightening a stranded container vessel by use of Chinook helicopters

High tiers of containers on deck could be out of reach of mobile cranes and subsequently awkward to remove, and in the past heavy duty helicopters have been given the task. Helicopters have associated operational problems, for example excessive turbulence, noise and vibration. Downdraft and the dangers of loose objects on deck may also cause injury to personnel.

Use of floating crane barge for removal of heavy lifts

The sheer weight of heavy lift cargoes can be problematical for a conventional vessel when attempting a lightening operation. Specialized craft fitted with lifting gear to accommodate the required safe working load may need to be called in to carry out single or specialized lifts. Provided the depth of water around the stricken vessel is adequate to float the barge at load draught, and provided the crane jib has the 'outreach', these craft can effect a practical solution to heavy problems.

Salvage operations, by the very nature of the beast, are high risk activities and when external services in the form of heavy lift cranes or helicopters are engaged costs can go extremely high. The salvor therefore needs to consider alternatives before involving third parties. The amount of the award would have to justify the high initial outlay for specialized equipment to effect recovery.

Towing and salvage

In the event of volunteering or being called upon to engage in a towing operation the Master should ensure that his/her own ship's charter party and bills of lading permit such an activity to take place. Once these facts are clarified, the practicalities of conducting a successful towage assignment need to take account of the following: assuming that the vessel requesting the tow is *not* in distress

(a) Has the towing vessel enough fuel/stores, etc. to complete the task.
(b) Confirm your actions with owners, underwriters and seek their agreement on your decisions. Insurance premiums will probably need to be increased.
(c) If under charter, the charterers need to be informed.

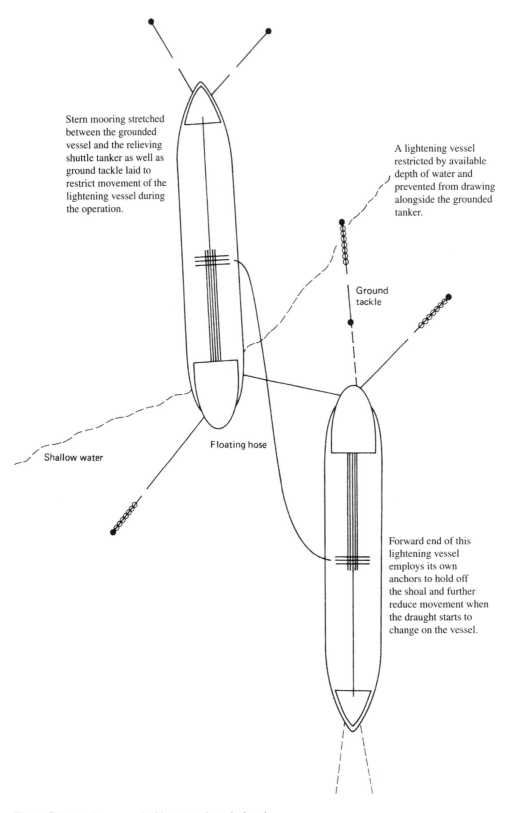

Stern mooring stretched between the grounded vessel and the relieving shuttle tanker as well as ground tackle laid to restrict movement of the lightening vessel during the operation.

A lightening vessel restricted by available depth of water and prevented from drawing alongside the grounded tanker.

Ground tackle

Floating hose

Shallow water

Forward end of this lightening vessel employs its own anchors to hold off the shoal and further reduce movement when the draught starts to change on the vessel.

Figure 5.11   Tanker aground with own anchors deployed

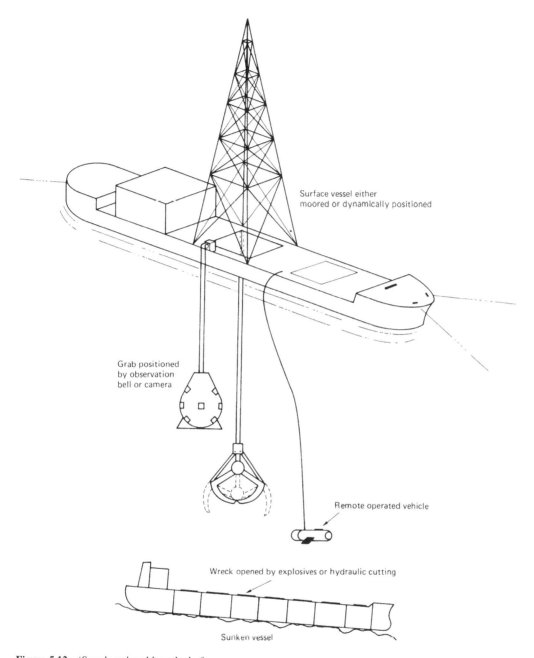

Figure 5.12 'Smash and grab' method of cargo recovery

(d) Are there risks to your own cargo, especially important with regard to perishables?

(e) Will the revised towing passage allow your own vessel to reach its loading port on time? If not, what penalties will be incurred and at what cost?

(f) Are the capabilities of your own vessel adequate to effect successful towing, e.g. main propulsion engines capable, deck machinery suitable, manning levels sufficient?

(g) Will the value of the towed ship and its cargo be worth the effort and expense incurred by the salvage operation?

(h) Can a salvage agreement be possible with the Master of the towed vessel, or will Lloyd's Open Form be established?

NB. The salvage award will depend on a successful completion of the tow and a Master should not hesitate to employ additional tugs at any stage of the passage to ensure success of the operation. However, any tugs so engaged should be hired on a basic 'contract of tow' and not engaged as a principal towing vessel.

It should also be realized that the above details reflect a situation where the towed vessel is not in distress. If a vessel is in distress, the obligations of the Master to assist a vessel in distress apply and would override all contractual terms in any carriage or insurance contract.

Figure 5.13   *Sea Empress* aground off Milford Haven, February 1996, attended by salvage tugs

Special circumstances

A derelict
In the event of a salvor taking possession of a derelict with no crew on board, and if he is the first to take possession, he will have absolute rights and control, provided it is a reasonable attempt to salvage the property.

Maritime property (for the purpose of salvage)
Maritime property is defined as any vessel used in navigation, together with its gear, cargo and wreckage from it, inclusive of rafts, boats, etc.

NB. It does not apply to lightships, buoys, navigation beacons or other similar sea marking devices.

Appointment/discharge of salvors
The Master of the vessel being assisted has the right to decide which vessel or vessels will provide the required assistance. Vessels that attend at a later time or interfere with the instructions of this Master could be sued for impeding a legal operation, and in any case would not be party to any award from the salvage claim.

Bimco Towcon
This towing contract was introduced in 1985 and has since received wide popularity in practice. The contract was drawn up by the Baltic and International Maritime Council, the European Tug Owners Association (ETA) and the International Salvage Union (ISU). It is recommended for use by all ISU and ETA members. The contract is widely employed where a lump sum agreement has been made for a specific towing operation. This lump sum is then separated into specific payments:

- amount payable on agreement and establishing the contract,
- amount due on the sailing of the tug and tow,
- amount due on passing specified landmarks,
- amount due on arrival at destination.

Account is also taken of bunkers and differences in costs within the contract.

Towing: precautions and safe practice

Any Master about to engage in a towing operation must evaluate the condition of the vessel being towed and compare the circumstances in relation to the proposed movement. As an example, a stranded vessel may have breached several compartments and any attempt to move the vessel before instigating repairs could result in the ship not surviving. It is essential that a detailed assessment of the towed vessel is made before commencing the operation and once agreement is established, the proposed passage plan must be investigated.

The passage plan (tug and tow)

The proposed route should be planned well in advance and take account of expected adverse weather conditions. Weather routeing advice should be employed where appropriate and available. The use of favourable currents, tidal streams and prevailing wind directions should be considered in preference to adverse elements.

Specific routes may be required under specific circumstances, such as a 'deep water route', if the underkeel clearance of the towed vessel is a restriction, or an 'ice-free route' might be required for a towing or towed vessel without ice strengthening.

Whatever route is finalized it must contain aspects of contingency planning. This would normally be expected to include precautions to compensate for the loss of the towline and/or the deployment of anchors, notably at critical positions on route where potential hazards would pose an additional threat.

NB. The towed vessel if manned should always have at least one anchor ready for use throughout the passage in case of emergency.

Prior communications to vessel traffic systems must be made to allow navigational warnings to be posted if and when the towing operation is passing through a controlled area, e.g. English Channel, MAREP and CNIS. Any special signals required by local by-laws additional to the normal towing displays need to be ascertained prior to departure and exhibited in ample time for specified areas.

Depending on the circumstances of the operation, a routine tow may be supervised by a towing master, the main function of whom would be to supervise the rigging and conduct the towing operation. The towing master would normally work in close liaison with the Master of the towing vessel and as such would be directly concerned with the passage plan and the proposed route.

Towing preparations

Whether engaging in routine towing or emergency towing consideration must be given to the following.

The vessel under tow

1. The vessel being towed must have watertight integrity throughout its length. All hatches and access points to the upper most continuous deck should be sealed and all watertight doors closed above and below decks.

2. The deck fitments, such as derricks or cranes, should be secured for adverse weather conditions, and loose objects stowed clear of upper decks.
3. The stability of the vessel should be calculated and the GM shown to be adequate. All tanks should be sounded prior to commencing the passage and any free surface effects reduced. Ideally, the towed vessel should be trimmed by the stern, with enough forward weight to avoid pounding.
4. The ship's rudder should be secured in the amidships position.
5. If the anchor cable is forming a part of the 'towline', the remaining anchor should be left in a state of readiness for use in an emergency.
6. Any overboard discharge valves should be closed and respective lines blanked off. The stern gland, rudder carrier and stuffing boxes should be tightened hard down.
7. Navigational signals for use by day and night need to be displayed respectively.
8. If the vessel is manned, the lifesaving appliances need to be sufficient to ensure the essential needs of the crew's safety are respected and the vessel is not deficient regarding survey requirements. Communications are essential and need to be tested before departure.
9. The amount of oil carried by the tow should be reduced to a minimum to offset the effects of any pollution which may arise following a mishap.
10. Emergency towing arrangements, together with means of boarding, should be readily available in the event of the towline parting and being lost. A trailing float line or alternative buoy recovery system should be available for locating and re-establishing the towline on such an occasion.

The towing vessel

The conduct of a towing operation should be supervised by a competent towmaster, who should not only establish an effective towline, but consider and influence all aspects of the intended passage.

1. The Marine Coastguard Agency should be advised prior to the voyage of any manned, towing operation.
2. The towing vessel should be established with correct navigational signals, for both day and night-time periods. Restricted in ability to manoeuvre lights and shapes should also be readily available.
3. The towing vessel must have suitable towing points and leads. Underdeck structure of bollards should be inspected to ensure adequate strength and if need

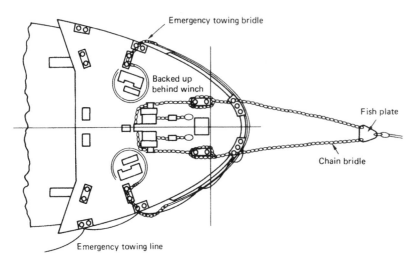

Figure 5.14   Purpose-built towing bridle arrangement

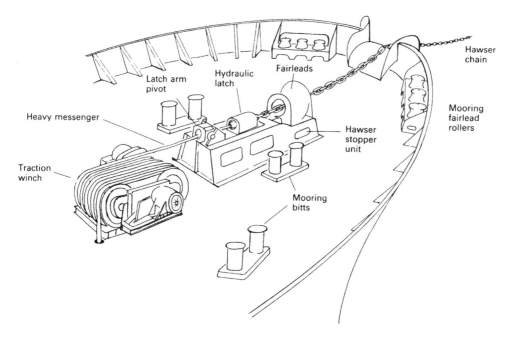

Figure 5.15   Specialist vessel using a towline via a centre lead

be additional towing brackets should be fitted, if possible.

4. A suitable towing speed for the passage and the prevailing weather conditions should be agreed between all parties.

5. Adjustment of the length of the towline and the position of setting the towline would be influenced on route by experience, depending on circumstances and sea conditions.

The detail and consideration given to a planned towing operation will depend on many factors, not least where the tow is departing from and where the operation is going to. Other considerations are whether the tow is manned or unmanned and the weather conditions.

Large towing operations may involve more than one single towing vessel, and the object being towed could be exposed above the surface or may be partially submerged. In any event, certain activities could pose a navigational hazard to other traffic and as such adequate advice should be passed to relevant authorities whose territorial areas may be affected, to allow the navigational warning service to be effective.

Towlines

Purpose-built towing bridle arrangement

Figure 5.14 shows a vessel prepared for long-distance towing where both anchors have been left operational. An additional chain is employed to construct the bridle arrangement around the aft part of the windlass using the ship's mooring leads and bitts. An emergency back-up towline is also set in place in the event of the primary bridle arrangement not holding.

As shown in Figure 5.15, specialist type vessels may be in a position to accept a towline via a centre lead and employ the vessel's existing mooring arrangements, e.g. tankers loading from FSUs.

Types and selection

The choice of towline to be employed in a towing operation could well be dictated by the circumstances and what materials are readily available. In any case a full evaluation of the vessel to be towed must be made before attempting to move her from the initial position. On occasion it may be prudent to move a vessel temporarily into a harbour where improved facilities to establish the towline could be better provided. Alternatively a Master could be faced with a situation of having to engage his own resources, for example a distress situation in open water.

Many smaller towing operations have been successfully achieved by a good mooring rope. However, larger and heavier towing operations have employed either a composite towline, utilizing the anchor cable of the vessel being towed, or bridle arrangements on one or both vessels. Tanker vessels over 20 000 GT are now equipped with Smit towing brackets (see Figure 5.16) fore and aft and these can be gainfully employed with appropriate towlines.

Preparations for a long-distance towing passage incorporate either an emergency recovery system in the event of the towline parting or an emergency back-up, towing arrangement by way of a second bridle, or a combination of both secondary rigs. Typical double arrangements are often required when operating in ice with ice breaker assistance.

The leads available for establishing towlines may not always present themselves on first inspection. Some ships are fitted with centre (bull ring) leads, while others may only be able to use the hawse pipe after first hanging off an anchor. A notable point is that some more recent towing operations secured towlines direct to the ganger length or to the anchor crown 'D' shackle, leaving the anchor attached. This avoided the work of hanging off the anchor and had the effect of damping down the towline under the surface and reduced sheering by the towed vessel.

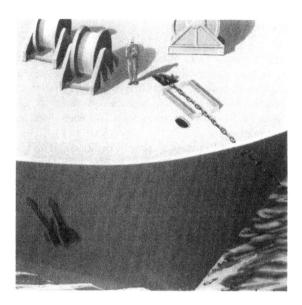

Figure 5.16 Smit bracket – tanker towing arrangement

Where a towing bridle is to be used a suitable size of chain to accommodate the weight and stress factors should be established together with appropriate anchor holding points. The ideal example is, of course, the strong point, used in conjunction with the tankers' Smit bracket (see Figure 5.17). Specific ship types may be more suitably fitted to engage in towing with a chafing chain coupled to a smaller hawser chain. Such an arrangement employs the principle of a bridle system, but care must be taken to ensure that the chafing chain extends beyond the lead and the joining arrangement is as strong or stronger than the chain elements, so as not to present a weak link in the towline.

Figure 5.17 Strong point and box stowage arrangement for chafing chain and messenger/towline leading through centre fairlead

The towage contract

There are many types of towage contract and they have a variety of clauses which may cover a precise service or range of services. Whatever type of towage contract is employed Masters should note that it is not a salvage contract but an ordinary contract which outlines a service for a payment.

The United Kingdom Standard Conditions (UKSTC) for towage and similar services include such terms as pushing, pulling, escorting, guiding, holding, moving or standing-by to describe its operational activities. As such defining 'towage' could only be considered in the broadest of terms. Under the law of the land, this flexibility of terms ensures that where liability is brought into question each case is judged on the facts and the individual circumstances of the case.

Onus of liability

The important feature of a towage operation has always been the relationship between the tug and the towed vessel. In the eyes of the law the tug and tow are recognized as one ship and the tug is the servant of the vessel being towed.

Where a conventional towage operation is being conducted, with a central command position aboard the towed vessel, it is usual for the towed vessel to be responsible. It is assumed that the towed vessel is vicariously liable for the actions of his servant, the tug.

However, this situation does not always follow, and the courts would seek evidence that the action of the tug is correct throughout, and that no third parties have influenced the outcome, before apportioning any liability. Initially it would be the responsibility of the tug owner to provide a tug in a seaworthy condition and having the capabilities to carry out the operation in the first place.

In the event that the tug owner/tugmaster fails to use his best endeavours, or is found guilty of negligence or misconduct, then they would be considered liable for any damage incurred.

Marine pollution

Every year over 3 million tonnes of oil are spilled into the marine environment, much of this from refineries. Less than one-third of this pollution originates from shipping casualties while another third is pumped out by tanker vessels cleaning tanks. Tighter regulations have in fact reduced the overall quantity released from tankers, partly influenced by the introduction of segregated ballast systems, but reduction is not the same as elimination.

The control of discharge of waste oils and garbage into the sea by ships is regulated by the International Convention for the Prevention of Pollution from Ships 1993, generally acknowledged as MARPOL 73/78, and the Merchant Shipping (Prevention of Oil Pollution) Regulations 1983 (SI 1983 No. 1398). MARPOL contains measures designed to prevent pollution caused through accident or routine shipboard operations. The regulations are inclusive of five annexes which deal with pollution by oil, noxious liquid substances, harmful packaged substances, sewage and garbage.

Oil pollution has probably been the most visible in the eyes of the general public since the *Torrey Canyon* disaster in 1967. With regard to oil (Regulation 16) the policy is such that the discharge of oil into the sea is limited and completely prohibited in specific 'special areas', e.g. Black Sea.*

One of the objectives of MARPOL was to effectively reduce the amount of oily water mixtures from operational oil pollution from ships. This was achieved for all ships of more than 400 GT, other than tankers which had restrictions placed on their activities when:

(a) Inside a special area.
(b) Within 12 miles of the nearest land.
(c) Other than proceeding on route.
(d) Oil content of any discharge had to be less than 100 ppm.
(e) An oil discharge and monitoring system had to be effective on board.

Additionally for Tanker vessels:

(a) The ship must be 50 miles from the nearest land.
(b) The instantaneous discharge rate must be not more than 60 litres/mile.
(c) Total discharge must not exceed 1/15 000 cargo for existing vessels and 1/30 000 cargo for new ships.
(d) The vessel must be fitted with an oil discharge monitoring and control system.

---

* Designated special areas include:
(Annex 1) – Oil.
Mediterranean Sea, Baltic Sea, Black Sea, Red Sea, the Arabian Gulf area, the Gulf of Aden and the Antarctic.

(Annex II) – Liquid chemicals.
The Baltic and Black Seas.

(Annex V) – Garbage.
Baltic Sea, Mediterranean Sea, Red Sea, Black Sea, Arabian Gulf area, North Sea, the Antarctic and the wider Caribbean region.

NB. Oil discharge monitoring and control system includes oil/water separator, oil filtering equipment or other similar installation.

Under the MARPOL regulations and guidelines, if a ship does not comply or presents what is considered an unreasonable threat then such vessel could be denied port entry. Penalties could also be imposed on the Master and owner of a ship which has illegally caused discharge. On conviction on indictment, an unlimited fine could be imposed.

Shipboard oil pollution emergency plan (SOPEP)

In order to ensure containment of a pollution incident, Masters and crews must have readily available for reference a SOPEP which details:

1. A procedure to be followed by the Master and crew to report an oil pollution incident.*
2. A list of persons or authorities to contact.
3. A detailed action plan to be taken up by persons on board to reduce and control the discharge of oil following an incident.
4. Designated procedures and point of contact for co-ordinating shipboard action with local and national authorities in combating the pollution.

Application of SOPEP

The Shipboard Oil Pollution Emergency Plan is applicable for any non-tanker vessel of 400 GT, or any tanker vessel of 150 GT and over. It must be in a format in accordance with the guidelines stipulated by IMO and be written in the working language of the Master and officers of the vessel.

The SOPEP could expect to include:

1. A procedure to be followed by the Master or officer in charge of a vessel reporting a pollution incident.
   WHEN: a discharge exceeds the MARPOL limits.
   a discharge to save life or property.

*The Marine Coastguard Agency of the UK has published the 'UK Safe Seas Guide' (wall chart displays sheets 1 and 2) (April 1998).
These display charts advise on When, Who and How to report a pollution incident around the UK coastline. Other information is detailed with regard to navigational hazards and areas which operate high speed craft on a regular basis, Marine nature reserves, areas to be avoided by shipping, pilotage requirements, weather services and marine communications relevant to safe navigational practice.

a discharge which has resulted from damage.
a threat of discharge is present.
HOW: (when at sea)
by the quickest method to a coast radio station, a designated ship movement reporting station, or an MRCC.
(when in port)
by the quickest available means to the local Harbour Authorities.
WHAT: An initial report of the incident, plus a follow-up report.
Characteristic detail of the oil spilled and cargo/ballast/bunker disposition.
A weather report, inclusive of sea conditions and any observed movement of the slick.

2. A list of authorities and/or contact persons in the event of an oil spill or pollution incident.
WHO: the nearest Coastal State when at sea.
the harbour/terminal authority when in port.
the ship's owners/manager, P & I insurer/correspondent.
the charterer and cargo owners.

3. A detailed description of the activity to be taken by persons on board the vessel involved in the incident to reduce and control the continued discharge of oil.
   In the event of a spillage caused by a casualty this should include:
   (a) the immediate actions required to preserve life and property,
   (b) the immediate actions to prevent the escalation of the incident,
   (c) information with regard to damage assessment procedures.
   In the case of operational spills:
   (a) any preventive measures and procedures,
   (b) any actions in the event of pipeline leakage, tank overflow, or hull leakage.
   A contact source should also be available within the manual for items of concern such as damage stability, stress considerations and calculations. Such contacts could include the Salvage Association, Lloyd's Register, or Classification Societies.

4. Procedures and designated points of contact aboard the vessel for co-ordinating any shipboard action with national and local authorities in containing and combating the spillage:
   (a) Actions required to initiate a local response.
   (b) The ship's responsibility to monitor clean-up activity.
   (c) Any assistance that the vessel itself can provide.
   (d) The details of any equipment or materials on board the vessel that could assist with deck clean-up operations.
   (e) Any details of local facilities and policies regarding a response to an oil spill.

Figure 5.18   Warship engaged in replenishment but restricted in its ability to manoeuvre

(f) Guidance with regard to maintaining records and sampling procedures.
5. Relevant shipboard documentation:
   (a) Ship's plans, diagrams, lists, e.g. ship's particulars, table of tank capacities.
   (b) Company's organizational chart.
   (c) List of key contacts.
   (d) List of company contacts, e.g. P & I Club and correspondent, company's agents.
   Additional material could include general arrangement plan, piping and pumping arrangement, cargo stowage plan, bunker details, damage stability criteria, shell expansion plan, list of shipboard clean-up materials, etc.

Spillage response

Any vessel, not just tankers, could become involved in a pollution incident through any one of several causes – collision, grounding, internal explosion, enemy action, etc. and how the Master and crew respond could influence the overall long-term effects considerably. No individual could expect to devise an action plan that would encompass every detail of every incident because each situation will be different and influenced by many variables the weather, type of vessel and geography, for example. However, a general approach to cover common needs to ensure essential activities is appropriate, hence the need for the ship's emergency plan contained in SOPEP.

Communications

In addition to the obvious distress or urgency signal that might be generated the Master should report the incident to the nearest Maritime Rescue Co-ordination Centre or sub-centre. Clearly the sooner the authorities know of the situation the sooner they can conduct an effective response employing all the available resources from shoreside facilities. Local authorities will need to be alerted to enable the community to participate decisively and for local amenities to be usefully deployed.

Damage/limitation/control

Any action taken by the Master to reduce the overboard discharge to effect containment within the vessel must be considered advantageous. For example:

(a) Listing the vessel to bring damaged areas above the waterline could eliminate or reduce the pressure in a damaged side tank.
(b) Sealing upper deck scuppers and freeing ports quickly, once a vessel runs aground, could prevent water pressure pushing the contents of double bottom tanks up sounding and air pipes onto the uppermost continuous deck.
(c) Establishing a collision patch over a damaged area.
(d) Internal transfer from damaged tanks into undamaged spaces.

Good seamanship practice is often achieved by improvization. Also, damage control activities depend on the ability to gain access to the damaged area – rarely possible on a vessel aground with bottom damage – and the immediate availability of appropriate resources on board.

Containment

Positive action must be taken once oil pollution has occurred and this could be in the form of extending communications once a damage assessment has been completed.

At the time of writing limited activity to contain a spill is available and falls into five main categories:

1. Barrier/containment boom equipment.
2. Skimmer vessels and pumping/collection facilities.
3. Lightening by external transfer into shuttle tanker or storage barge.
4. Dispersal chemicals usually deployed by air.
5. Burning off.

In order to effect any of the above a Master needs to bring in technical assistance to achieve containment. The sooner instructions are in place the sooner specialist parties can participate in recovery/clean-up operations.

Not all Port Authorities have the necessary facilities to manage a large spill, and without doubt would need to bring in additional skimmers, pumps, dispersal chemicals, technicians and specialized labour, air support, lightening vessels, etc. Even if the Harbour Authority had enough to cover a basic emergency, a large spill would require an excess of resources.

Case study – lessons from the *Exxon Valdez*, Prince William Sound, Alaska (March 1989)

The worst oil spill in North American history released some 11 million gallons of Alaskan crude oil only 25 miles from the Valdez oil terminal. The contingency plan for such an accident was instigated and the Port of Valdez was closed. Unfortunately, the plan was woefully inadequate for such a large spill. Clean-up equipment was not readily available in sufficient quantities and was slow to move into position. Skimmers were initially deployed on the 4 inch black surface scum, but only five were available and didn't have the capacity required and proved almost useless against a 100 square mile slick. See Figure 5.19(a) and (b) for a comparison of the growth of oil slick over 24 hours.

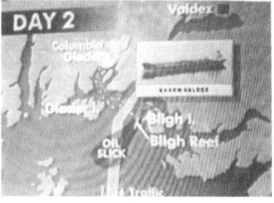

(a)

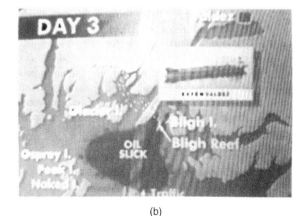

(b)

Figure 5.19 The oil slick from the *Exxon Valdez*, grounded at Bligh Reef, expands to eventually cover 100 square miles

Dispersal chemicals were found to be non-effective on oil that had been in the water for several days and aircraft deploying the chemicals had some difficulty in sighting and spreading on the slick because of the black waters of the region. A combination of delayed chemicals and discussions on the long-term worsening damage to the ecology further slowed down the operation.

Surface containment booms were deployed over several miles, but received criticism as 'too little, too late', and even when in position considerable seepage was noted from the barrier equipment. Igniting the oil and burning off was found to be reasonably successful but not practical for such an enormous area because of the considerable air pollution associated with this method.

Clean-up ashore

With any oil spillage, large or small, comes the clean-up operation, usually organized by the local authority and very often supported by voluntary groups and individuals. The extensive use of manual labour in virtually every case must be anticipated and they can expect to employ a variety of equipment from the basic shovel to heavy plant bulldozers, steam hoses or high pressure water jets.

Steam or high pressure water hoses have proved useful in cleaning oil off rocky surfaces. Once cleansed from the rocks the oil drops back into the sea and can be removed by skimming operations. Absorbents can be used with some success and have been found to be useful in and around inaccessible places where manual methods are difficult to pursue.

Bioremediation has been used in some spillage areas where phosphorous and nitrogen nutrients were applied to spilled oil to generate the growth of naturally occurring oil degrading bacteria. However, this is slow in relation to the volume moved by other manual methods.

Varieties of wildlife always tend to suffer from a spillage. Estimates of damage to fish stocks, sea birds, mammals, shell fish are beyond costings in dollars. It was estimated that half a million sea birds, thousands of mammals, including whales, seals and sea otters, died in the *Exxon Valdez* disaster. Salmon, crab and cod stocks suffered drastically, which had a subsequent effect on other areas of the fishing industry.

US oil pollution act 1990 (OPA '90)

As a direct result of the Alaskan oil spill the United States Congress amended the Federal Water Pollution Control Act/Clean Water Act and incorporated the following provisions:

(a) An Oil Spill Liability Trust Fund of $1 billion was provided for states to have access to combat any future disaster. Funding for this came from increased taxes on crude oil and recovered penalties.

(b) The authority for a Federal On-Scene Commander (FOSC) was established for the control of oil spills and clean-up operations.
(c) A National Strike Force Co-ordination Centre (NSFCC) was established.

The work of the Marine Pollution Control Unit (MPCU)

The Marine Pollution Control Unit is part of the Marine Coastguard Agency, which is an executive agency of the Department of Transport. Its function is to take action when a pollution incident by oil or other dangerous substance occurs around the UK. (Many countries worldwide have their own emergency response units.)

Based in Southampton, the MPCU generates a response from the emergency operations room, following an assessment of the situation based on information and data received. Their action is directly related to the casualty which may require a search and rescue operation. In the event of pollution it has been realized that oil clean-up at sea is not always an option before environmental damage has occurred. For this reason priority becomes the prevention of further leakage.

Salvage operators may be contracted in to transfer the cargo to a support craft, or tugs might be brought in to assist the casualty. The MPCU has the authority to carry out such operations as necessary to reduce and if possible eliminate the pollution risk to the environment by whatever means.

NB. The clean-up operation on shore is primarily a local consideration for the Port and Harbour Authority, which should have contingency plans. However, the MPCU would provide advice on planning and support of local authorities and would also assist in clean-up activities in the event of a major spill.

The work of the Marine Pollution Control Unit includes the conducting of training courses for local authority operatives and emergency planners. They also carry out Research and Development programmes to improve counter-pollution measures and are actively engaged in pollution prevention measures, i.e. air patrol surveillance.

The MPCU maintains the National Contingency Plan which contains the arrangements for dealing with spillages of oil or other hazardous substances into the marine environment. They participate in formulating international agreements and would also co-operate internationally where any incident involved another country.

Merchant shipping regulations (prevention of pollution)

Initial survey

An initial survey of the ship is required to ensure that the vessel complies with all the requirements of the International Oil Pollution Prevention Certificate (IOPP).

The survey will consist of a thorough examination of:

(a) all ship's plans, specifications and technical data relating to the design of the ship and its equipment to ensure that they meet the requirements of relevant regulations,
(b) all manuals, Oil Record Books, certificates and other relevant documents, and that they are confirmed on board,
(c) the ship's condition and all equipment, ensuring that it is constructed and installed in a correct manner to satisfy regulation requirements.

Annual survey

This survey should in general consist of the following:

(a) a visual examination of the vessel and its equipment, together with its certificates. Certain tests may be carried out to confirm correct maintenance procedures are being followed.
(b) a visual examination of the ship and its equipment to ensure that no unapproved modifications have been made which could be detrimental to working operations.

Intermediate survey

This survey is a thorough examination held at $2\frac{1}{2}$-year intervals ± 6 months. It should be extensive enough to satisfy the surveyor that the pumping and piping system, inclusive of the oil discharge monitoring and control equipment, complies with the regulation requirements, and effectively remains in good working order.

The Oil Record Book

Every ship must be provided with an Oil Record Book (Part 1 – Machinery Space Operations) and every oil tanker must be provided with an Oil Record Book (Part 2 – Cargo & Ballast Operations). Entries must be made on a tank to tank basis, if appropriate, whenever there is a movement of oil inside or outside of the vessel's hull.

NB. See *Seamanship Techniques*, Part 2, for further information regarding example entries.

The Master must sign each page and the Operation's Officer must sign for each specific operation. The book(s) must be held on board and be ready for inspection at any reasonable time. Such records must be retained for a period of 3 years.

The Master should also be aware that any person so authorized may inspect the Oil Record Book(s) and may also take certified copies from the Master for use in judicial proceedings.

Future changes towards tanker safety

It has been realized for some time that older ships are more likely to be involved in pollution incidents than the more up-to-date tonnage. One of the possible reasons behind this was that vessels built over 20 years ago were not forced to comply with the more tighter standards that have been imposed on the newer building of today. The requirement for double hulls is a prime example which is being phased in over a period of time.

NB. New oil tankers must now be constructed with double hulls and existing tankers must be converted when they become 25 years old. At the time of writing approximately 300 of the world's 3500 tankers are fitted with double hulls. Clearly such a small number of vessels could not transport more than a very small percentage of the world's oil, neither do the existing shipyards have the capacity to carry out total mass conversions. It therefore follows that a level of practicality must enter the safe seas equation.

The International Convention on Oil Pollution Preparedness, Response and Co-operation (OPRC) was adopted in 1990. It was designed to facilitate international co-operation in the event of a major oil pollution incident. Oil pollution co-ordination centres have since been established at IMO headquarters (1991) and in Malta for the Mediterranean Sea, and an information centre and training facility has been established in the Caribbean (1996).

Certificates and documents summary

The following certificates and documents are required by merchant vessels under MARPOL and associated annexes:

• International Oil Pollution Prevention Certificate (IOPP) – valid 5 years, may be extended by 5 months.

- Oil Pollution Insurance Certificate (OPIC) – valid 1 year and cannot be extended.
- Shipboard Oil Pollution Emergency Plan (SOPEP).
- Oil Record Book.
- Certificate of Fitness (for chemical and gas carriers) – valid 5 years and cannot be extended.
- International Pollution Prevention Certificate for the Carriage of Noxious Liquid Substances in Bulk (INLS) – valid 5 years.
- Cargo Record Book.
- International Air Pollution Prevention Certificate (IAPP).
- International Sewage Pollution Prevention Certificate (ISPP).
- Garbage Record Book.
- Garbage management plan.

NB. UK Oil Pollution Prevention Certificate (UKOPP) is not required by MARPOL.

Bunkering procedures

Not every pollution incident is caused by a major accident like grounding or collision. Very often a spill could be the result of taking bunkers on board, which must be considered a routine practice for any vessel. However, the effects of complacency, poor training or unforeseen circumstances could find the vessel and the Master involved in an unwanted pollution incident.

The usual practice aboard vessels is that Chief Engineers take delivery of bunkers, but the security of the deck falls to the responsibility initially of the Chief Officer and ultimately the Master. Under the ISM code, designated responsibilities are identified to ranking personnel under the company umbrella and specific safety checklists accompany all bunkering activity. The Master needs to ensure that Chief Officers follow the checklist policy which should include the following items.

Prior to taking bunkers

1. Seal up all deck scuppers to prevent spillage overside.
2. Establish a second means of access to the vessel in case of a fire outbreak obstructing the main gangway.
3. Display appropriate signals 'B' flag, or all round red light.
4. Post additional 'No Smoking' signs in exposed deck areas.
5. Establish full fire precautions close to the manifold and ensure immediate readiness and availability.

6. Set up and test communications between the pumping station, the manifold and the reception personnel monitoring the delivery.
7. Make sure adequate drip trays are positioned under flanges and in way of the manifold.
8. Rig fire wires, fore and aft, if appropriate.
9. Have dispersal chemicals readily available for use on board the vessel in the event of spillage.
10. Detail sufficient manpower on deck and in the engine room to carry out the operation correctly, especially when 'topping up'.

All preparations should be noted in the log book and heads of departments should be advised of the intended activity.

Engineers participating in any bunker operation are expected to be familiar with the relevant tanks, vents and valves involved. Valve systems to tanks not being used should be securely shut and safeguarded against opening or operation. A full set of all tank soundings should be obtained prior to the commencement of loading, including tanks due to be loaded. Valves on the 'fill delivery line' should be seen to be open before pumps are engaged to commence at an initial slow delivery rate.

On loading bunkers

1. Ensure communications are maintained throughout and that monitoring of tank levels is continuous.
2. Ensure that the 'No Smoking' policy is adhered to.
3. Tend moorings and gangways as the draught of the vessel is increased.
4. Establish a continuous watch over the manifold/pipeline to ensure ship movements do not incur a bad lead or fracture on the loading pipe.
5. Have the emergency contact numbers to hand, for use in the event of spillage.
6. Tanks, when filling, should be identified as such on the level indicator.
7. Once pumping has commenced, the Master and Chief Engineer should be kept informed of the operation's progress.
8. Additional manpower may need to be kept on standby and brought into the operation for 'topping up' or in the event of a spillage.
9. The Harbour Authority should be kept informed throughout up to and including completion times.
10. Once loading is complete and the pipe is landed, tank soundings should be confirmed and an entry made into the Oil Record Book.

NB. Chief Officers should note the ship's draughts on completion and are expected to carry out a stability check prior to sailing. Changes to the ship's ballast tanks may be required if loaded bunker quantities are unequal to port and starboard fuel tanks.

Management of garbage

Pollution can occur in many different forms and oil pollution especially receives a very high profile. Yet pollution of the environment by garbage is just as prolific but has not attracted a similar level of attention.

The Prevention of Pollution 'Garbage Regulations' 1988 (SI 1988 No. 2292) applies to all UK ships, fishing vessels, offshore installations and yachts. The application of these regulations is effective on UK vessels wherever they may be and is also applicable to other non-UK vessels when inside the territorial waters of the United Kingdom. (For additional reference see MARPOL 73/98 and the 1995 amendments, Annex V.)

The definition of 'garbage' is any victual, domestic or operational waste, which is generated by the usual operation of the vessel, which is liable to be disposed over a continuous or periodic period. Exceptions to this definition are fresh fish or parts thereof, and sewage (the latter is covered by MARPOL Annex IV).

Since the 1995 amendments to the convention, every vessel over 400 GT, and every ship certificated to carry 15 persons or more, must now carry a garbage management plan (GMP).

This plan must outline the procedures for the collection, storage, processing and disposal of on-board garbage. It will also contain details of equipment on board that is employed to accommodate the handling of generated garbage.

The type of equipment required must satisfy the regulations of specific areas and will normally be in the form of grinding and/or comminuting garbage sufficiently to allow the waste to pass through a screen/sieve, with openings not greater than 25 millimetres.

Shipping is clearly not the only culprit that generates garbage. A great deal is washed up on beaches from people ashore, for example, holidaymakers who leave rubbish behind when they leave the beach. Other sources are fishermen, or local councils who dump garbage into rivers or directly into the sea. In order to protect the environment there is an ongoing need to educate people and authorities not to use the oceans as a giant rubbish tip.

One problem within the shipping arena has recently been accentuated by the growth in the cruise/passenger market in and around the Caribbean. The wider Caribbean area attracts considerable cruise traffic and the sheer volume of garbage generated from one cruise ship is considerable, but multiplied by 20 or 30 ships of a similar class then the garbage problem becomes a cause for concern.

The itinerary of a cruise vessel is such that many visit one port a day, and few ports have effective facilities for dealing with garbage. The fact that not all have ratified Annex V of MARPOL, for one reason or another, does not help towards a policy of cleaner seas. If shipping companies and/or ships themselves contributed more financially to establishing reception facilities then not only could the environment be better protected, but ships would have one less problem to contend with.

Various authorities have issued local advice and the ship's Master should be familiar with the environmental protection regulations effective for specific areas. An example of this is the UK Safe Seas Guide (wall poster) issued in April 1998 by the Marine Coastguard Agency. Mariners are advised as to what is and what is not permitted in the North Sea, Irish Sea and the Atlantic Ocean and what are considered the minimum ranges from land where certain classes of garbage may be disposed of.

For additional references on garbage disposal regulations, see the Merchant Shipping (Prevention of Pollution by Garbage) Regulations 1998 (SO) and the Merchant Shipping (Port Waste Reception Facilities) Regulations 1997 (SI 3018 No. 1997) (SO).

Garbage categories

Disposal outside special areas

Material	Category	Conditions of disposal
Plastics	1	Must not be disposed in any region.
Dunnage, packaging material	2	Flotable, may be disposed of 25 nautical miles or more from the nearest land.
Food waste, not ground or comminuted, inclusive of rags, paper products, glass, metal, crockery, bottles and similar	3 and 5	May be disposed of 12 nautical miles or more from the nearest land.
Food waste ground or comminuted and all other ground or comminuted waste	4 and 5	May be disposed of 3 nautical miles from the nearest land.

Disposal inside special areas

Special areas	Condition of disposal
Antarctica, North Sea, Arabian Gulf, Red Sea, Black Sea, Baltic Sea, Mediterranean Sea, English Channel and wider Caribbean Sea	No garbage whatsoever other than food wastes may be disposed of.

Disposal from offshore installations

Disposal of garbage from a platform or from a ship alongside or within 500 metres of such a position is prohibited, except where the platform is more than 12 nautical miles from the nearest land, and food waste is ground or comminuted.

Breaching the regulations

Any owner, ship's manager or Master may be subjected to an unlimited fine if the garbage regulations are breached. The defence against such a charge would mean that proving the discharge was necessary to save life, or was caused by damage to the ship or equipment and that all reasonable precautions were taken to minimize the escape to the environment.

Garbage record book

In accord with the regulations, vessels are expected to keep and maintain a record of any treatment of garbage in the form of a record book. Such records should include:

(a) When garbage is discharged into the sea.
(b) When garbage is deposited at reception facilities ashore or passed to other ships.
(c) When garbage is incinerated.
(d) When accidental discharge of garbage takes place.

Each entry into the records must stipulate the time that any garbage is disposed of and the position of the vessel at that time. Each incident of handling garbage must carry the signature of the officer in charge of the operation and specify the circumstances, the type and the amount in cubic metres being disposed of. The Master would normally ensure that all movement of garbage is controlled and would sign each page of the record book.

For additional reference see MSN 1720 (M & F).

Heavy lift operations

Cargo work has always been the domain of the ship's Chief Officer, and the conventional heavy lift by crane or derrick is an associated activity. With the heavy duty derricks of the Stulken variety, and reference to the manufacturer's/rigging plans, most Chief Officers cope satisfactorily. As with any heavy lift operation the Master should be kept informed. However, with weights in excess of 600 or 700 tonnes, a Master should be involved, especially on the stability factors, without interfering with an efficient officer.

Stability factors

With a conventional vessel it is normal practice to calculate the maximum angle of heel that will occur during the lifting period, and a junior officer monitoring the 'inclinometer' once the lift is commenced could give early warning if this heel angle is being exceeded. (In such circumstances the load would be landed and the stability reworked.)

With reference to Figure 5.20 Masters and senior officers should be conscious of the fact that if GG1 is increased, the angle of list will increase. Whereas if GM is increased, the angle of list will decrease.

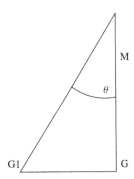

Figure 5.20

In order to achieve a successful outcome an adequate GM must be established prior to commencing the lift, bearing in mind that the gravity effect of the load will act from the head of the derrick once the weight is taken on the lifting gear. Prudent preparations should include an assessment of the vessel's ballast tanks and the elimination of any excess free surface effects. Additional ballast may be usefully added low down to improve the ship's GM

Lateral drag

One of the main problems with a vessel landing a heavy lift and carrying the expected list will be the dangers of the lift dragging itself from the loading platform as the weight is relieved and the ship rolls back to the upright. This situation and associated dangers can be eliminated by coming back on the lifting purchase and simultaneously coming back on the topping lift, so keeping the derrick head in a plumb line.

For an example of an offshore side heavy lift using a Stulken double pendulum derrick, see Figure 5.21.

Figure 5.21 Example of a heavy lift carried out on the offshore side by a Stulken double pendulum derrick – a ram's horn lifting hook being employed with a four point bridle and spreader arrangement

Floating sheer leg operations

Figure 5.22 shows equipment used in a heavy lift operation. Taklift 4 has a maximum lift capacity of 2400 tonnes operating with the 'A' frame boom (54.4 m) and works with a 30 metre 'fly-jib' to a capacity of 1600 tonnes.

Heavy lift vessels: float on-float off principle

Exceptionally heavy loads, often in excess of several thousand tonnes, tend to be lifted and transported by the principle of 'float on-float off' by specifically designed heavy lift ships or barge carriers (see Figures 5.23, 5.24 and 5.25).

Merchant ship entry into war zones

In the current world climate, a Master could very well find that his ship is heading for a port inside an area of ongoing hostilities or a declared war zone and as such some guidance regarding ship preparation might be considered useful.

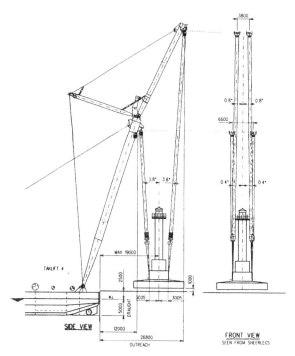

Figure 5.22 Taklift 4 of Smit Transport and Heavy Lift shown lifting navigational markers for fine positioning between Malmo, Sweden and Copenhagen

Figure 5.23 Example of heavy lift transport: tanking system fitted low down below the loading deck and equipped with large capacity pumps. Load area designed substantially with underdeck strengthening and additional pillar support

Prior to sailing

A Master should consider the passage plan and keep out of the range of shore-based military hardware, and if possible stay clear of hostile territorial waters for as long as possible. The Master should also advise the company of the intended route, agree communication times and positions beforehand, and provide the company with a

Figure 5.24 Similar heavy lift vessel loaded after floating on a jack-up, offshore installation drilling unit. The vessel is ballasted down to bring the loading deck below the waterline permitting the heavy lift load (usually rafted) to be floated on board. Once in position over the load deck the ballast tanks are pumped to raise the load deck and the heavy lift over and above the waterline

(a)

(b)

Figure 5.25 Barge carrier. Similar high capacity tanking system but gains flexibility by being fitted with a heavy duty lifting crane to provide load on as well as float on capability

comprehensive list of crew, their next of kin, addresses and contacts.

Full bunkers, fresh water and general stores should be taken on, and the catering officer should be advised to take on suitable foodstuffs, which could be useful for injured parties (soups and liquid products), as well as tinned goods that could be doubled as emergency rations in survival craft.

Consideration might be given to carrying out modifications to the vessel's structure; not necessarily major changes, but providing steel shutters for the bridge windows or establishing a heli-deck operational area might prove beneficial once inside a war zone.

When it is known beforehand that the ship's destination is inside a hostile area a prudent Master should consider ordering additional and what must be considered essential stores:

1. Medical supplies should be enhanced with extra casualty/injury supplies, e.g. burn dressings, splints, wound dressings, bandages, morphine, etc.
2. The number of fire extinguisher refills, and spare extinguishers, spare B/A cylinders, fire suits, and similar protective clothing should be increased.
3. Damage control materials such as timber supports, collision patch materials, steel plate, extra cement, anti-pollution equipment, blackout curtain material, etc., should be taken on board.
4. Engine room spares, extra fuses, light bulbs, pump parts, oil filters, gaskets, torch batteries, etc.

On route

While the ship is on route, the Master should consider increasing the training activities of the crew, especially in the use of essential elements like survival craft, fire-fighting activity and medical practice/casualty handling.

Prudent action should also include extensive preparation across all departments to ensure the following:

1. Examine and check the quantity and condition of lifeboat stores, and add any useful additions, e.g. blankets, water containers, cigarettes, playing cards.
2. Establish a damage control store, with fire-fighting supplies, emergency portable lights, extension cables, crowbars, large hammers, sand, cement, timber, torch batteries, first aid kits, protective clothing, etc.
3. Rig lifelines, and check the rigging of embarkation ladders, lifeboat painters, accommodation ladders, lifebuoys.
4. Ensure spare lifejackets are in place at the bridge, engine room and damage control centre.
5. Test emergency systems: steering, lighting, emergency generator, lifeboat engines, emergency fire pumps, communications, alarm systems and watertight door operations.
6. Clean out and prepare hospital accommodation with casualty handling supplies. Include spare clothing and additional blankets.
7. Stow away and secure all loose and non-essential small gear.
8. Establish an emergency navigation station external to the bridge, with back-up facilities by way of charts, pencils, parallel rules, nautical tables, sextant, spare compass, binoculars, etc.
9. Inspect the vessel thoroughly with particular attention to watertight integrity of weather deck entrances, correct positions of fire-fighting equipment, emergency storage lockers and contents, hospital condition, galley condition, and ensure alleyways are not obstructed. Also the securing of survival craft.
10. Conduct a crew briefing on survival actions in the event of abandonment, surface attack and aircraft attacks.

On approach to the area (avoiding potential hazards)

The Master should be aware that if a vessel is struck by explosives or incendiary projectiles, anything inflammable on board the vessel could feed the ensuing fire. To this end all curtains, not being used for 'darkening ship' purposes, should be stripped and stowed, together with alleyway carpets, rugs, mats and fabric upholstered furniture. Similarly all pictures, lamp shades, paper on notice boards, etc., should be cleared away.

Glassware is extremely hazardous, and windows should be cross-taped and if possible masked. Glass shelving, mirrors, bottles, drinking glasses, etc. should be boxed and stowed into secure cupboards.

Ship's furniture, if non-essential, should be stowed clear of working areas or lashed out of the way. Tables and chairs that can be secured by location chains should be fixed tightly. Any soft furnishings which are likely to give off toxic fumes should be removed from the normal working environment. All portholes/scuttles should be secured and deadlights positioned. Fire doors and watertight doors should be closed and only opened briefly for access when necessary.

Inside a hostile area

Different circumstances could and will influence the Master's decisions and an assessment of the risk factor regarding the level of hostility towards commercial shipping will be needed. For example, could the vessel be subject to missile attack (as seen in the Gulf conflict in 1994, and in the Falklands War in 1982).

Whatever the degree of threat will dictate the precautions and safeguards the Master must order. However, if the threat does not materialize nothing is lost. A typical order may be to ask personnel to shave off beards. Crew should also be discouraged from wearing man-made fabrics, to avoid the dangers of the material melting on the skin.

Some people may consider the above instructions to personnel as an infringement of their civil liberties, but if it saves an arm or a face scarred for life, then it is fully justified.

Once inside a hostile area, the safety of the ship as well as its personnel become paramount and it would be negligent of the Master not to have the vessel at a state of readiness prior to entering such an area. The bridge team should be at an alert status, with double lookouts posted during the hours of daylight and darkness. Communications should be kept to a minimum and radio silence may be necessary.

The charter party: General War Clause

In the event of the vessel entering a war zone or the nation under whose flag the vessel sails is engaged in warfare then the Master should note the content and any restrictions that may be placed on the ship by what is described as the 'General War Clause' within the charter party. Different charter parties will word the clause relevant to the specific ship or trade but in general it would be concerned with the safe navigation of the vessel and the well-being of the crew and cargo.

Masters are usually advised not to sign bills of lading for ports which are blockaded. They should note what alternative actions are required in the event of a port becoming blockaded after loading has been completed and where bills of lading have already been signed. Options on cancellation of the contract would usually also be stipulated.

Because of the proximity of hostilities the future movements of the ship would be considered in this clause. In most cases the ship would have the liberty to comply with instructions from the government of the flag under which the vessel sails regarding routes, ports of call, arrival and departure destinations, etc.

The nature of the cargo in transit could influence any decision by the Master, especially if the cargo is liable to be classed as contraband of war, or runs the risk of being confiscated. Such concerns may well involve direct communications with cargo owners as well as the vessel's owners regarding acceptable destinations for discharge.

In the event that the vessel is ordered to an alternative port, it is not usually taken as 'deviation' and the vessel is considered to have fulfilled the voyage contract. However, Master's would be well advised to keep detailed records of all communications and instructions if presented with such circumstances.

War zone: Master's concerns and duties

Unless a Master is in communication with his own military forces it would be wise to attempt to maintain a level of secrecy with regard to the ship's position, especially important if carrying war supplies as a cargo. An enemy would not be adverse in attempting to prevent war materials reaching their destination by whatever means possible.

Nothing should be thrown overboard from the vessel on approach to or inside a war zone, especially garbage. Rubbish should be incinerated wherever possible and if it must be dumped, then only at night and in weighted containers. Tell-tale garbage trails have given away many a ships' position to a sharp-eyed aircraft pilot. Good housekeeping amongst crew, and comprehensive briefings by alert officers, can be life saving in the long term.

No one can predict the future and outcome of a voyage and from the outset it is the Master's responsibility to think ahead towards the unpleasant task of taking casualties when inside a combat zone. The crew might be well advised to make a will before departure in the event that the vessel comes under fire and sustains casualties.

The Master should also tend to any classified documents or equipment which should be prepared for destruction before entering any hostile area. The fact that his ship could come under attack is a fact of life and it is wise to be prepared if the vessel is to be boarded.

Summary

Few people willingly go into a combat area, and even fewer would go in unprepared. With a little foresight Masters could reduce the obvious risk to ship and personnel by seeking advice and taking some basic precautions. Some of the items discussed may seem far-fetched to those without experience of a hostile zone. But who decides what is too much and what is not enough? Once a Master leaves a war zone without a full crew, the question must be asked, did I do enough?

The risk of piracy

Introduction

Piracy is described as an act of robbery and/or violence on the high seas. It is usually carried out under arms and has frequently involved murder. It is strongly associated with armed robbery which is committed inside territorial waters and under the jurisdiction of the flag state.

There is currently an increase in both piracy and armed robberies against ships in various parts of the world, notably South-East Asia, and Central and South America. See Figure 5.26 for a map showing historical areas of piracy attacks an shipping.

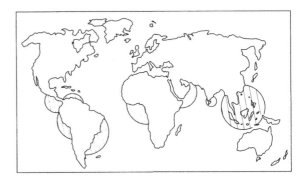

Figure 5.26   Historical areas of piracy attacks on shipping

The indicated areas provide an overview of where predominant incidents of piracy continue to occur. However, Masters should be aware that attacks are not limited to these areas and isolated activity could and does occur elsewhere. Although not as common, incidents have been recorded in open sea conditions, but the extended range and vastly improved communication networks tend to deter this type of attack because of the increased chance of detection and subsequent apprehension of the culprits.

The modern 'pirate' has changed from the earlier image of a Captain Hook and often employs intelligence before planning and then attacking a ship. An attack will often employ high speed, small craft and the attackers are often armed with a variety of weapons including automatic fire arms.

Attacking methods

Methods of attack on ships vary considerably depending on the circumstances and the position in which the vessel finds herself. Ships at anchor are extremely vulnerable and can be easily boarded from small boats under cover of darkness. Combined attacks on vessels that are tied up alongside are also common, for example, seemingly innocent parties engaged on board the ship as stevedores are in fact found to be dropping stolen goods overside into waiting, motorized canoes – a method employed extensively in Central American regions, where the culprits can make a speedy getaway up or down a river, away from the ship's berth.

In the areas of Indonesia, Thailand, the Malacca Straits and the South China Sea a common practice is where fast launches pace the target vessel, which is often underway and making way at quite a fast speed. During the hours of darkness these craft close the ship to within grapple range while both craft are at speed, at which point the ship is boarded. Boarding in this way is not without considerable risk to the individual and is made easier with vessels having a low freeboard.

One of the main objectives of a boarding party is to obtain control of the ship's main operational areas such as the bridge and the engine room. By intimidation of crew, and especially the Master, boarders can then engage in the theft of cash, ship's equipment, parcels of cargo and personnel effects. On occasions, it has also been the practice for boarding to take place with stealth in mind, to force the Master to open the safe without the alarm being raised and make an equally discreet escape without raising the attention of the crew. This method may possibly reflect some inside knowledge of the ship's internal arrangement.

Piracy targets

The majority of pirate attacks are to secure cash and easily transportable valuables and to this end the Master's/ship's safe is a prime target. This is especially so if it is known that the vessel is carrying a lot of currency, for example crew wages. Alternatively, it might be known that the vessel is carrying a particular, small, valuable cargo, like bullion, and this could lead to the vessel being singled out for attack.

The so-called 'soft target' must be considered the slow vessel, with low freeboard and a small crew. The ship with low level security (possibly influenced by reduced manning), poor deck lighting and poor communications equipment could well influence the choice of target to be attacked.

Prior knowledge of the vessel and its route would allow forward planning by pirates. Inside knowledge of the ship's internal arrangement, location of the safe or cargo parcels, crews quarters, etc., all help a successful attack. The method of attack, where, when and how would no longer become a matter of chance, but a well-planned, positively executed and a surprise operation.

Ship's defence against piracy attacks

Any vessel which must pass close to or through an area where pirates are known to be active should take some precautions in order to deter potential attacks. The prudent Master could formulate a defensive stance against incident to include an anti-attack plan and brief his officers and crew accordingly. Consultation with the ship's owners to gain support for the installation of surveillance equipment should also be sought. Such items as deck and access cameras (CCTV) to monitor approaches to specific on-board operational areas, coupled with an alarm system, extra search lights and back-up communication systems must all be considered as active deterrents towards preventing an attack.

Crew training with briefings on what response to make in the event of a potential attack should be an aspect of a ship's readiness. Deck patrols, increased lookout duties and establishing secure areas should all form part of routine procedures when in high risk areas. (For additional reference, see MGN 75 (M) – this Guidance Notice states that training should be based on the fact that an attack will take place, and not in the belief that with good luck it would not.)

Up-to-date information on piracy incidents should be sought from the local authorities of the area concerned and can also be obtained from the ICC International Maritime Bureau's Piracy Reporting Centre, Kuala Lumpur, Malaysia (Tel. +603 201 0014, Fax +603 238 5769, Email ccskl@imbkl.po.my).

In the event of attack

Once an attack is underway, the Master should raise the alarm by radio, as in many areas surface and helicopter patrols are conducted by the Coastguard and Military Services. These may be within operational range and could deter an attack from being successful. However, not every vessel will be successful in evading an attack and may experience pirates eventually boarding the ship. The undesirable situation may be thwarted by prudent use of radar to detect small craft approaching. Once their intention is clear the crew switch on all deck lights and train jet water hoses to prevent boarders reaching the deck. The crew must be warned in such a case not to become an exposed target against deck lighting in case the would-be attackers are armed.

The ship's speed should be maintained and abrupt helm movements might be possible to create a wash against small hostile craft, provided the navigation circumstances permit such hard-over actions of the helm.

Once boarded the main concern must be for the safety of personnel on the ship, and retaining control of the navigation of the vessel. If the alarm has been raised the attackers would probably be anxious to clear the vessel as soon as possible, and nothing should be done in any way to obstruct their departure.

Each incident will of course be different, but it is not unusual for crew members to be seized and used as hostages to intimidate the Master to open the safe or supply specialist information on valuables. Incidents have occurred where the whole crew have been seized and locked up. The attackers took what they wanted and departed the ship leaving the vessel without effective control. Clearly this state of affairs could have led to a collision or other similar accident and possible pollution of the area.

NB. The use of hoses as means of preventing hostile craft closing the ship are recommended at a pressure of about 80 lb per square inch. This level of pressure if directed correctly would probably prevent a man from standing upright. The water could also cause the outboard engines to short out as well as cause swamping of the boats themselves.

Once on board, attackers should not be antagonized by bravado tactics which could not possibly succeed against an armed party, and only result in injury to innocent personnel. Far better to encourage the attackers to depart the ship as soon as practical. Aggressive actions by the Master or Officers could well result in reprisals, and should not be advised unless there is a 100 per cent chance of success, without incurring casualties. (Highly unlikely if the boarding party is large and also armed with fire arms.)

The use of firearms

Masters may feel that better protection can be achieved by issuing firearms to crews for personal protection and for the defence of the ship as a whole. It should be pointed out that personnel would need some specialized training for the use of firearms and such weapons would in themselves become an added target for pirates.

The British authorities discourage the use of such weapons on the basis that they could escalate an already highly dangerous situation. It should also be realized that if a foreign national is killed or injured, even in self-defence, this could generate repercussions on the ship and its crew from a localized flag state.

Jurisdiction

Armed robbery within the territorial waters of a flag state falls to the jurisdiction of the Coastal State where the offence is committed, whereas piracy is a criminal act on the high seas and outside the jurisdiction of any flag state. The jurisdiction which then applies to any pirate apprehended comes under the flag state of the captor of such a person.

Intervention by warships

Under international law any warship would be expected to take action to suppress an act of piracy and come to the aid of a vessel being attacked on the high seas. However, foreign warships engaged in territorial waters of another nation cannot exercise force or pursue attackers without express permission from the Coastal State. An exception to this would be in a case of distress or where a vessel found herself in danger, when the warship could assist on humanitarian grounds.

Generally world navies have seen a reduction in the numbers of ships and it is unlikely that a commercial vessel will have the benefit of its own nationality's 'man of war' in close proximity when an act of piracy is ongoing. Masters should therefore be aware that they could be isolated for some time before any help is forthcoming.

Example

Singapore anti sea robbery measures
Operated by the Police Coastguard, these measures cover acts of 'sea robbery' inside territorial waters. Daily patrols by the Police Coastguard and vessels of the Republic of Singapore Navy are conducted in the area and extensive bilateral communication links are established with the co-operation of the Marine Police of Malaysia and Indonesia in combating this type of crime.

Case history examples

February 1996, fishing vessel MN-3
Normina – located in the Southern Philippines
This vessel was engaged in fishing with a crew of ten, when two speed boats approached. Both the speed boats carried two men each armed with automatic weapons. As the craft approached the fishing boat, they opened fire killing nine fishermen and injuring the tenth man. Despite head injuries the injured man escaped by swimming away from the vessel, while the attackers made themselves busy attaching towlines to the fishing vessel.

June 1996, MT Bow Fortune – located at anchor off the Port of Salvador, Brazil
At 0145 hours, on 26 June 1996, the *Bow Fortune* was boarded on the port side of the main deck by seven

armed men. One watchman was struck with the butt of a pistol as he tried to run away and a second watchman was later captured.

The attackers forced their way to the bridge and held the Duty Officer and the watchmen at gun point for about 20 minutes. The Master's office was then broken into and the safe emptied. The attackers then proceeded to ransack all cabins and steal cash and valuables from the crew.

The whole crew were then assembled, their hands tied and were ordered into a kneeling position, while their rings and other jewellery were subsequently removed. Crew members were kicked and punched and some received injuries from pistol butts.

The attack lasted nearly 2 hours, the pirates leaving the vessel at 0340 hours.

Incident report

The International Maritime Bureau reported that in November 1998 pirates attacked the Panama flagged ship MV *Cheung Son*, a 16 785 dwt bulk carrier in the China Sea. All 23 crew members were murdered by machine gun fire after being lined up on deck. Their bodies were then weighted and thrown overboard.

Local fishermen found six bodies in their nets, still bound and gagged. Chinese authorities later arrested seven suspects, who confessed to the shooting and dispersal of the bodies.

This incident reflects the increasingly violent nature of attacks and the willingness of pirates to use automatic weapons to capture cargoes, and in some cases the whole ship. Although this vessel was boarded in the Taiwan Strait, vessels are extremely vulnerable when lying at anchor, passing close in shore of coastlines, or navigating a passage through islands known to be active with pirates.

Statistics

The list below shows reported incidents of piracy in 1998.

Indonesia	59
Philippines	15
India	12
Malaysia	10
Bangladesh	9
Somalia	9
Ecuador	9
Brazil	9
Kenya	7
South China Sea	5
Cameroon	5
Columbia	4
Tanzania	3

Future trends to combat piracy

In 1997, the IMO received reports on 252 acts of piracy which represented an increase of about 26 per cent on the previous year. In October of 1998, the start of a series of missions and seminars commenced with the objective of reducing the number of incidents by promoting regional co-operation. A seminar at Singapore in February 1999 was attended by expert representatives from such countries as China, Indonesia, Malaysia, and Vietnam, IMO Associate as well as Members, with the aim of increasing awareness and intensifying efforts to deal with the problems of armed robbery and piracy.*

It was noted that although the Malacca Strait and the South China Sea remain as major areas of concern, incidents in the Indian Ocean had increased from 30 in 1996 to 41 in 1997, while East Africa had increased from seven incidents to 11 in the same period. Other areas also showed increases:

West Africa from 21 (1996) to 30 (1997)
South America from 32 (1996) to 45 (1997)
Mediterranean/Black Sea from 4 (1996) to 11 (1997)

IMO have adopted a resolution, 'Measures to prevent acts of Piracy and Armed Robbery, against ships', which urges respective governments to take all measures to suppress acts of piracy or armed robbery against vessels in or adjacent to their territorial waters.†

It is also expected that all future incidents are reported with as much detail as possible, with the view to establishing patterns and methods of operation to enable countermeasures to be developed. One of the more recent ideas is a tracking device placed on ships which can be activated when a vessel is threatened. This would allow the local authorities to respond quicker and as such may establish prevention or even capture of the culprits.

Another positive move in the fight against criminal acts was the establishment of a rapid response team in 1998. The team would fly out to the next port of call of the vessel involved. The objective here is to obtain detailed information from the crew which can be passed to law enforcement agencies of the country where the incident took place. The view being that investigation by the agencies can commence sooner rather than later, with hopefully faster and better results than have been experienced in the past.

---

* For the year 1999 estimated losses due to piracy ran out at 100 million US dollars. Prevention funding declined and incidents increased during this same period.

† Some companies are now carrying ex-military security guards aboard vessels trading in high risk piracy areas of Southeast Asia.

6 Ship handling and manoeuvring aids

Introduction

Historically it has always been the domain of the Master to conduct the handling and manoeuvring of the vessel and rarely will another officer become directly involved. This state of affairs has been the practice for many years and does nothing for the training of the budding Chief Officer seeking command who will be expected to become the most experienced man afloat overnight.

It is understandable for a Master to be reluctant to allow another to handle the vessel in close confined waters, but with more sea room junior officers should be encouraged to advance their knowledge and skills for the future. Various texts, including this one, can expand on the theory and virtues of ship handling practice, but there can be no substitute for the 'hands on' operation.

Figure 6.1 The *Wellington Star* berthed alongside. Forward moorings well spread and taut (four inshore headlines and a bight, one offshore headline plus two backsprings). Starboard anchor deployed and cable in the up and down position. Accommodation ladder landed from all aft superstructure

Factors affecting ship performance

The speed of a vessel can only be maintained if the hull form is retained in a condition to offer least resistance to the motion. Such resistance can be increased by numerous factors including marine growth below the water line, modifications or damage to the hull, and corrosion being allowed to accelerate.

Periodic dry docking of the vessel allows overside maintenance to be carried out, permitting growth to be cleaned off and anti-fouling paint to be applied. At the same time similar activity can take place where corrosion is noticeable. Layers of rust tend to build up on the hull and touching up with paint can result in a mixture of proud areas and bare patches producing an uneven finish. Shot blasting or sand or high pressure water blasting techniques can clean large hull areas relatively quickly in the perimeter of a dry dock.

Damage to the hull, although not seemingly major, will also occur on a continuous basis while berthing. Small indents into the hull form will eventually produce an uneven surface and affect the water flow about the lines of the vessel. Additional resistance will result, causing reduced performance.

Not all water resisting factors are adverse and some are often deliberately incorporated in a trade-off for increased performance in other areas. An example of this could be where a propeller is retrofitted with ducting. Like the propeller this could increase the known 'drag effect' but the trade-off could be reduced cavitation and improved fuel/speed cost effectiveness.

Other modifications such as a bulbous bow are often incorporated to decrease the size of the bow wave and reduce the water resistance. However, it should be realized that the additional wet surface area of a bulbous bow will cause greater frictional forces to be present, the theory being that reduced wave resistance is more substantial than the increased frictional forces.

Some shipboard elements like bow thrust or stern thrust units, whether fitted at the building stage or added as a modification at a later date, are known to positively increase water resistance. This becomes acceptable where greater manoeuvrability and cost-effective savings can be achieved by reducing the need for tugs in berthing operations.

Ship trials and manoeuvring data

Prior to a shipowner accepting delivery of a new vessel the builders will be required to demonstrate the standard of the ship during performance trials. A Master taking over a new vessel could well be asked to join other observers from such organizations as underwriters, the Marine Coastguard Agency and/or representatives from the Classification Society to monitor and assess such trials.

Checks and tests

1. Tests on anchors and cables are usually carried out in shallow water and the length of time taken to bring the anchor 'home' is noted.
2. The windlass performance is monitored during (1) and the efficiency of the machine assessed.
3. Stopping distances under normal conditions are assessed.
4. Stopping distances under emergency conditions are assessed.
5. Speed trials which could be timed runs over a given distance are carried out.
6. Turning circles at various speeds are conducted.
7. Routine helm movements are ordered and movements checked against the helm indicator, the rudder and the rudder indicator.
8. Emergency steering gear arrangements are also tested to satisfaction.
9. Deceleration trials assess the time taken to stop the vessel from various speeds.
10. Emergency crash stop assessment from full ahead to full astern is checked. (Also checked at other speeds.)
11. All emergency shutdown systems are tested and checked for operation.
12. Navigational equipment is tested and assessed.
13. Fuel consumption for distance and speed comparisons is measured and usually monitored over a period of voyage time.

Results are recorded and charted and a copy normally remains on the bridge for direct access by the Master and ship's officers.

Deck preparations for arrival/departure

The most regular of shipboard activities occurs when arriving to berth alongside or to let go and take the vessel off the berth. This procedure has been carried out so many times one could be forgiven for assuming that it would never go wrong, even with experienced officers and men. Unfortunately, the operations do go awry very

often through complacency, forgetfulness or overlapping circumstances. The minor task of getting heaving lines out and ready could easily cause a delay and allow the wind to affect the vessel's position. The proverb 'for want of a nail the shoe was lost, for want of a shoe the horse was lost, for want of a horse the rider was lost' comes to mind.

This is not to say that the Master of the vessel is expected to prepare heaving lines and dot all the i's and cross all the t's, but the Master should think ahead and brief his station officers in ample time to ensure that stations are called early enough to provide adequate preparation time at the mooring positions. Special mooring operations discussed beforehand can ensure a smooth and unambiguous operation, and may save the day.

Communications should be tested well in advance, and station officers should ensure that their areas of responsibility are fully operational. Any defects in equipment or problems in manning must be reported to the Master earlier rather than later, to allow remedial action to be taken or amendments made to the docking plan, e.g. engaging manual steering and testing astern motion on engines well in advance.

The control of external parties such as tugs, pilots, linesmen, mooring boats, etc. are often out of the control of the Master and therefore due allowance and early booking through Port Authorities/company agents should be the order of the day. Confirmation of the same should be made at an appropriate time to ensure no unwanted, delays. If confirmation is made early enough and delays are expected, then the vessel's speed could be adjusted and ETAs revised.

Docking and undocking operations will vary from berth to berth. Some quays will not be well fended while others may require slip wires, buoy moorings, or the use of anchors. Different ports have local practices and for the Master new to the area, it would be prudent to ascertain as many details as possible by consulting the sailing directions/charts, communicating with pilots or Port Authorities, or discussing with own ship's officers with past experience.

All berthing activities are recorded in the Deck Log Book, and previous old log books could be a valuable source of information when a Master is docking in a port for the first time, or, of course, if the Master is newly appointed.

Checklist for arrival in port

1. Prepare an approach plan to include pilotage well in advance and consider use of the following: large-scale charts and port plans, sailing directions, tidal and current data, projected weather reports, and any available port information such as local by-laws.

2. Conduct a thorough check on the planned tracks to ensure that the underkeel clearance is adequate throughout the passage. Draughts and trim should be accounted for, which may require adjustment of cargo or ballast, especially if operating speeds are likely to cause encumbrance through 'squat'.

3. Sight and note any navigation warnings affecting the area of operation and ensure all working charts are up to date.

4. Test steering gear and carry navigational equipment checks prior to entering enclosed waters. Engines should be tested in astern motion.

5. Open up communications in accordance with the local regulations and pass on ETA together with details of dangerous or hazardous cargoes/goods carried and/or passenger numbers, etc.

6. Order pilot, tugs, linesmen, mooring boat and any other relevant berthing requirements.

7. Confirm communication channels and report in to any VTS system in operation.

8. Establish pilot rendezvous and boarding arrangements, local weather detail, vessel's heading and which side to create a 'lee' (assuming launch delivery). Delivery of the pilot by helicopter would require separate preparations as per the ICS guide 'Helicopter Operations at Sea'.

9. Engage manual steering and conduct operational checks on deck equipment, such as: windlass, anchors and cables, mooring winches and leads, mooring ropes and wires prepared, internal communications to fore and aft stations, signalling apparatus and deck lighting.

10. Warn crew departments in advance of standby, pilot boarding time, stations, berthing time and once Customs have cleared the vessel.

Clarify berthing details: use of anchors, gangway, shore connections, etc., as soon as practicable.

NB. Some vessels specifically equipped will need to 'house fin stabilizers' and retract recording logs in ample time, before entering narrow/shallow waters.

Mooring deck arrangement

Figure 6.2 shows the open fore deck of a passenger vehicle ferry. The figure clearly shows the centre line 'bull ring' with 'international roller fairleads' and 'Panama leads' set either side; split windlass with separate power/control units for cable and gypsy arrangements, tension drums and warping drums attached; two additional tension winches positioned either side, aft of windlass positions; bitts (bollard sets) strategically situated, to suit the lead positions; and the forward hatch leading to rope store.

Figure 6.2 Open fore deck of passenger/vehicle ferry

Ship handling: equipment and control

With the pace of new design and increased technology, ships are being fitted with more and more manoeuvring equipment. The bridge has tended to remain as the main control 'conning' position, but even this, on occasions, can be removed to a remote station. The modern bridge is now usually open plan and design has been generated by ergonomics and the specialist's needs of the trade. Bridge wing controls to port and starboard are now in common use in addition to the main control consul (see Figure 6.3). They are usually inclusive of pitch angle controls for controllable pitch propellers, thruster controls, stabilizer control units, steering and auto helm unit (usually on the fore and aft line). Depending on the size of the ship and type of trade it is engaged in, certain vessels may be fitted with doppler radars, rate of turn indicators and other sensing/monitoring instrumentation.

The modern vessel with the integrated bridge is designed to suit navigational needs as well as manoeuvring requirements and communication units, navigation light sentinels, depth recorders, speed logs and the like, all of which are incorporated to meet operational needs. The move towards open plan has also moved the chart room into the bridge space providing the navigational Watch Officer with continuity for lookout duty as well as position monitoring and other similar watchkeeping duties.

Current ship building activity is meeting the needs of an increased cruise ship market and the fast, high speed ferry traffic. Bridges are being optimized with 360° all round viewing, a requirement which might be considered essential aboard a high speed craft moving in excess of 50–60 knots. Other building reflects the specialized operations market with increased numbers of offshore standby/supply vessels or diving support vessels employing dynamic positioning control systems.

Figure 6.3 Enclosed bridge wing situated either side in an overside position and fitted with control consul to port and starboard. Main engine and thruster controls, fore and aft communications together with gyro repeater/bearing site vane are also visible

With all the changes taking place it is easy to forget the conventional ship, of which there are still a great number in operation. These vessels have been and continue to be regular workhorses for deep sea trades and are usually without bridge control consuls, stabilizers or thruster units. The skill of the Master with only a right-hand fixed blade propeller, basic rudder assembly and a marine diesel engine is paramount in achieving berthing objectives, and ship handling in these circumstances should in no way be seen as an easy option.

Modern rudders: manoeuvring

The need to improve efficiency when manoeuvring continues to be an ongoing process and the application of the type of rudder assembly employed will define the eventual outcome of any ship movement. From around 1975 the Schilling rudder became a noticeable advance in improving the turning ability of a vessel. Earlier, the use of rudder flaps and power rotors had evolved and these subsequently combined to provide the tighter turning circle.

Vectwin rudders (Schilling)

This system provides better course stability in comparison to a conventional rudder construction. It was appealing because unlike other rudder assemblies, i.e. the 'mariner rudder', there were no moving parts underwater which lent itself to longer service and reliability (see Figure 6.4).

Vectwin rudders have been fitted with rotary vane steering units and can operate with a single fixed pitch

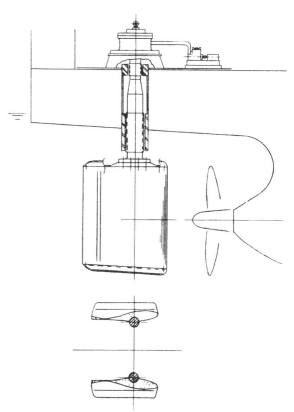

Figure 6.4 Arrangement of Vectwin rudders

propeller. The manufacturers claim the use of the system equals that of a controllable pitch propeller (CPP) when considering speed control. Savings could then be achieved by shipowners at the building stage by alternatively installing Vectwin rudders as opposed to the more expensive installation of a CPP.

From the point of view of a Watchkeeping Officer, a single joystick control would ensure easy manoeuvrability for the ship – the vessel responding to the direction in which the joystick is moved and the propulsive thrust being proportional to how far the joystick is moved from the hover position. Bridge instrumentation provides visual feedback on rudder angles throughout any manoeuvre.

A powerful braking effect can also be achieved by this system and the amount of headreach a vessel would normally carry could be drastically reduced. It must also be anticipated that any lateral deviation of the vessel would be virtually eliminated when compared to, say, an engine put into an astern movement, as when taking all way off.

When the turning circle is considered, the rapid speed reduction caused by larger rudder angles of 65° or 70° results in lesser angles of heel than those incurred by a conventional rudder. Also, because speed is reduced,

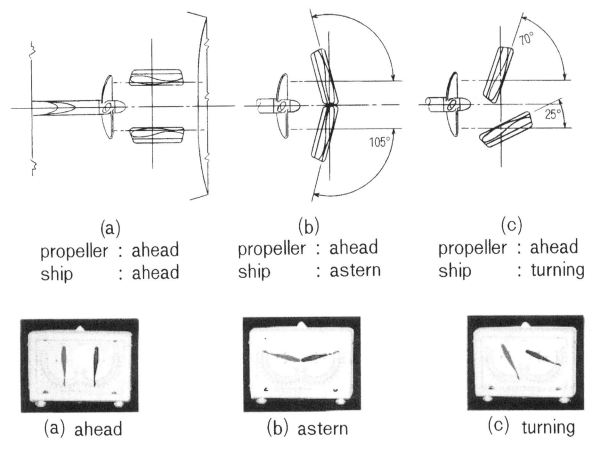

(a)	(b)	(c)
propeller : ahead	propeller : ahead	propeller : ahead
ship : ahead	ship : astern	ship : turning

(a) ahead   (b) astern   (c) turning

Figure 6.5 Typical rudder angles for basic manoeuvres

'advance' and 'transfer' are considerably less and the subsequent turning ability of the vessel is expected to be a lot tighter than a conventional ship (see Figure 6.5).

The Becker rudder

The Becker rudder is a two-part design, with a hinged flap connected to the main body of the rudder. When causing movement of the main rudder, the direction of the flap is altered by mechanical linkage to approximately double the angle of deflection placed on the body of the main rudder (see Figure 6.6).

The Becker designs can be fully supported with a heel pintle bearing arrangement set into the 'sole piece' of the ship's stern frame. Alternative designs include a semi-spade or a full-spade option, where the lower part of the rudder is left unsupported (see Figures 6.7 and 6.8).

Figure 6.9 shows the Becker, king support rudder. This type of rudder is secured to the hull by means of a rudder trunk with an inside bore to accept the rudder stock. The end part of the rudder trunk accommodates

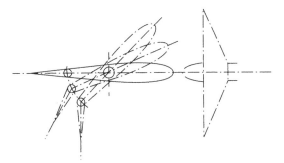

Figure 6.6 The resultant combination of both rudder and flap produces a strongly curved 'aerofoil' section generating a lift increasing effect

an inside bearing for mounting the stock while an outer bearing supports the rudder blade.

An example of a Becker flap rudder can be seen in Figure 6.10. All the moving parts and those turning against each other are manufactured in stainless chromium–nickel steel, or plastic (polyamide, nylon),

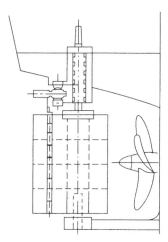

Figure 6.7 Fully supported rudder

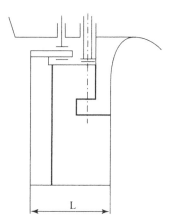

Figure 6.8 Semi-spade rudder

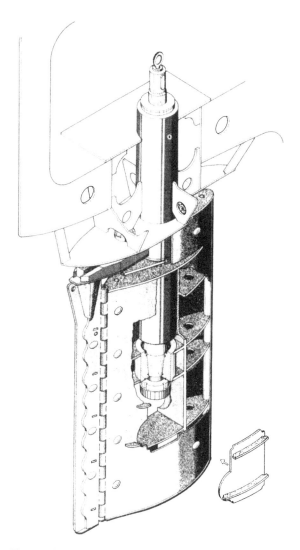

Figure 6.9 The Becker, king-support rudder

material. The use of such materials has distinct advantages in that no lubrication problems become apparent because all bearing points are in touch with water. The bearings are smooth in operation with a low frictional coefficient, while the 'bushes' have a high wearing resistance.

The bodywork of the rudders is welded steel sheet plate and they are manufactured to meet classification standards. In the event of damage occurring to the flap by way of the mechanical linkage failing or the flap is taken off, the rudder would still function as a conventional standard rudder but with diminished manoeuvrability.

Rudder response

In tests and in practice the Becker rudder reflected between 60 and 80 per cent higher transverse forces when compared with similar forces on a conventional

style rudder. A tighter flow was achieved at angles of up to 30° onto the low pressure side of the rudder which also provided a desirable underpressure effect. While at a larger 45° angle with the flap at 90° a generally superior turning effect was achieved in comparison with rudders of similar areas set at similar angles.

The drag effect of any rudder is notable, however, with the action of the flap at small angles of, say, 1° or 2° – the deflection is more immediate because of the double deflection which affects the ship's ahead course more quickly, while at the same time not increasing the drag effect. Practically, because the Becker design only requires about half the rudder angle of a conventional design the drag effects are correspondingly reduced. This means that an improved speed can be realized by the vessel.

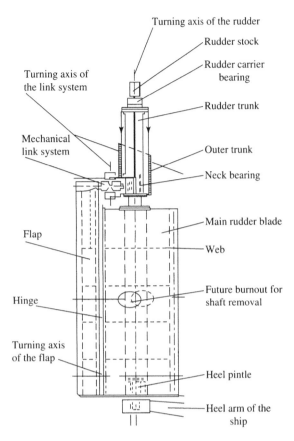

Turning axis of the rudder
Rudder stock
Rudder carrier bearing
Turning axis of the link system
Rudder trunk
Mechanical link system
Outer trunk
Neck bearing
Flap
Main rudder blade
Web
Hinge
Future burnout for shaft removal
Turning axis of the flap
Heel pintle
Heel arm of the ship

Figure 6.10   Design of Becker flap rudder

A vessel fitted with the Becker rudder and a bow thrust unit could expect to obtain an extremely tight turning ability. This could in turn reduce the need to employ tugs in areas where vessels fitted with the more conventional rudders would find the need to engage tug assistance essential.

Summary

The ability to be cost effective in ship management is now a fact of life and any aids that can reflect speed/fuel economy or a reduction in overheads by reducing or eliminating tug use will obviously attract keen attention. From the Master's point of view, as the ship handler any improved equipment that can be engaged, especially without having to change operational practice, and at the same time improve performance, has to be welcomed.

Arguments for ducted propellers

Ducted propellers have been employed for over 50 years, and mariners continue to be divided on the merits when compared with open propellers. Many ships are now operational with ducted propellers and the cost of fitting requires some justification.

An immediate advantage at the building stage is the fact that ducted propellers are generally smaller than open propellers. This in itself can produce savings in both the working and any spare propeller carried by the vessel. Size reduction also provides a limited weight saving benefit, albeit small. Against these factors are the additional cost of 'ducting' and the subsequent maintenance once in service.

From the ship handler's perspective ducting would appear to suppress propeller excited vibration, especially that generated from cavitation. Tests have indicated that in general, a vessel fitted with a ducted propeller, when in the loaded condition, experienced less vibration at the stern and at the position of the thrust block, than a sister ship which operated with an open propeller. However, when in ballast, the ducted ship was observed to experience twice the levels of vibration at the stern than the vessel fitted with the conventional propeller. In conclusion, provided the vessel is gainfully employed and operating in a fully loaded condition with continuous cargoes, lesser levels of vibration will be experienced by the vessel.

It is now generally accepted that worthwhile gains can also be achieved in the ship's speed with vessels fitted with ducted propellers. Model tests together with full scale assessments have indicated enhanced speeds of up to as much as 0.6 of a knot. This of course could be translated into obvious fuel/power savings on those voyages where a payback period for installation and maintenance costs could be realized. Another way of applying the saving could be to offer a lower freight rate which could be attractive to shippers and provide a competitive edge.

Practical considerations should also take account of the possible damage incurred by the duct and/or the rudder or propeller. The duct could be repaired by conventional shipbuilding methods but the propeller could only be withdrawn if space permits, without disturbing the rudder. A new ship could of course be built to accommodate this need, but where ducts are fitted retrospectively this lack of clearance could be a problem in the event of such damage (see Figure 6.11).

Conclusion

Ducted propellers have seemingly proved their economic worth, although they have not been without teething problems such as corrosion and have experienced more problems when retrofitted as opposed to being fitted when the vessel is built. It should therefore be an obvious consideration in the early stages of design for future tonnage.

Figure 6.11 Fully ducted propeller arrangement and 'spade' type rudder under dry dock maintenance operation. Attention is drawn to the close proximity of the rudder's forward edge to the ducting surrounding the propeller

Use of thrusters

Many vessels are currently fitted with either bow or stern thrusters (see Figure 6.12), or a combination of both. However, it is not necessarily the type of ship but more often the 'trade' that the vessel is engaged on which dictates the need for thruster units. The nature of a trade which requires a vessel docking two or three times a day would clearly benefit from the fitting of thruster units to reduce the need for tug use, e.g. ferries.

Another obvious advantage of a vessel fitted with thrusters is increased manoeuvrability. A ship equipped with thrusters experiences better handling characteristics when at slower speeds and improved safety levels could be obtained when turning in adverse weather – a faster turn effectively reducing the risks from beam seas. Maintaining head to sea as when 'hove to' could also

Figure 6.12 Triple bow thrust tunnel units under maintenance on a cable laying vessel while in dry dock

be achieved with relatively few problems being anticipated.

Specialist type vessels like rig support vessels, survey or diving support craft clearly benefit from thrusters. Many of these offshore working vessels are now employing dynamic positioning (DP) where station holding by thruster units is the order of the day.

Although the initial cost of installing bow and stern thrusters may seem high the saving in towage costs from not having to engage tugs can be substantial.

Example of towage rates during 1998:

Small tug	with 10–15 tonnes bollard pull	$350	per hour
Medium tug	with 15–25 tonnes bollard pull	$1200	per hour
Large tug	with 25–35 tonnes bollard pull	$2500	per hour

The fact that towage companies usually operate a 2 or 3 hour minimum hire charge tends to put pressure on Masters to minimize the tug requirements. Based on the above figures it would not be unreasonable to expect a bow thrust unit to be paid for from the savings of tug costs over a 2 to 3 year period.

Additional savings are also achieved by many Masters with well-equipped vessels engaging in their own pilotage operations where thrusters can make the task much easier. A common practice on the regular liner/ferry trade routes.

For an illustration of the arrangement and control of a tunnel bow thruster, see Figure 6.13.

Effectiveness of thrusters

A great percentage of ship handling is carried out when the vessel is stationary and if the thruster is engaged at this time it would be expected to produce the design output. However, in practice it is often seen to be most effective when the vessel has sternway. The reason for this is that the pivot point of the vessel is set approximately one-third of the ship's length from aft. When the vessel is making headway the thruster would tend to have little effect if the speed is over 4 knots. The pivot point would also be established one-third of the ship's length (approx.) from the bow and the thruster effect would be reduced in this instance.

It is unusual to find thrusters fitted in isolation, and they are more commonly employed as part of the overall design configuration. The accompaniment of controllable pitch propellers (CPP) and improved rudder designs is the accepted norm, especially for vessels with high manoeuvring demands.

Configuration examples

150 000 tonne shuttle tanker:
Twin controllable pitch propellers + split twin rudders + 2000 hp bow thruster.

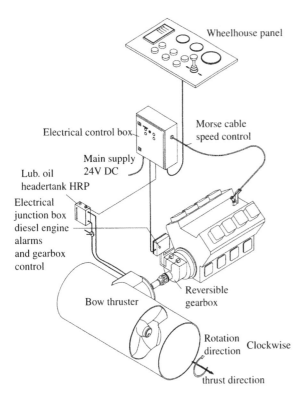

Figure 6.13   Arrangement and control of a tunnel bow thruster

Joystick control can link the above to achieve dynamic positioning capability.

Large container vessel:
Controllable pitch propeller + Schilling rudder + 1800 hp thruster.
Capable of berthing without tugs in up to force 5 or 6 winds.
(A typical thruster unit could expect to turn a 20 000 tonne container vessel at approximately 70° per minute.)

Types of thrusters

Tunnel thrusters are probably the most common, being a propeller mounted in a transverse tunnel inset at the bow or stern areas. Earlier types were of a fixed blade propeller, shaft driven by either a diesel or electric motor. More recent designs have employed Controllable pitch propellers with electric supply.

Jet thrusters draw water from an external source by means of an impeller and eject either a large volume of water at low speed or a small volume at high pressure, causing a reactionary thrust in any direction, e.g. white gill systems – see *Seamanship Techniques*, Part II.

Cycloid thrusters (Voith Schneider)

These are a suspended, movable blade system, turned by a constant speed motor. The resultant thrust is produced in a single direction, and may be forwards, astern or to port or starboard. The direction and the thrust power developed can be controlled by the operator. These are popular with harbour craft, tugs and the smaller type vessel. (See Figures 6.14 to 6.16.)

Thruster control and operation

Figure 6.17 is an example of the control and operation of a Kort thruster.

Azimuth thrusters (steerable thrusters, omnidirectional or azipods)

These are probably the most useful of all the marine handling equipment currently being integrated into modern tonnage (see Figure 6.19). They are essential for DP operations with diving support vessels and other offshore work boats, and are extremely popular with vessels which require high manoeuvrability, e.g. tugs, harbour craft, supply boats, etc.

The majority of steerable thrusters turn through 360°, are usually ducted and may have shrouds. However, a recent building the *Botnica*, on long-term charter with the Finnish Maritime Administration, has been constructed with the dual role of ice breaker in the winter and offshore supply vessel in the summer. This vessel is equipped with twin azipods (external motor/propulsion units) each being power rated with 5 mW, and devoid of nozzles. Trial work showed that the azipods were more effective without nozzles when in ice conditions.

Thrusters can often be employed in isolation as a main propulsion system eliminating the need for a rudder assembly, or as an addition to a conventional machinery/rudder design. Azimuth thrusters are either fixed blade or controllable pitch propellers. If the latter, the angle of pitch can be reversed or the thrusters can be rotated through 180° to change direction. (Depending on the manufacturer rotation of the thrusters takes approximately 0.14 seconds.)

Example of the thruster use

An automated system fitted with three azimuth thrusters and joystick control to cause a changing thrust from ahead to astern is achieved by one thruster turning through 180° while the other two outer thrusters reverse pitch (reverse thrust by changing pitch is 80 per cent

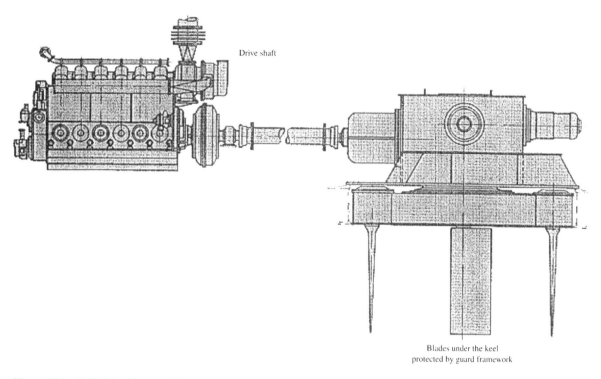

Drive shaft

Blades under the keel
protected by guard framework

Figure 6.14   Voith Schneider propulsion

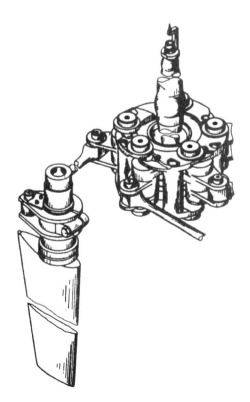

Figure 6.15   Gearing and linkage to a single blade

of the maximum ahead thrust). If astern power is still required after 60 seconds, a second thruster would rotate and if a continued requirement is needed, the third thruster would rotate.

The majority of azimuth thrusters are fitted with two operating speeds, 50 and 100 per cent.

NB. The BP tanker *Seillean* is fitted with six 3000 kW controllable pitch, steerable thrusters. The cost of such fitting would probably preclude such design for all but specialized vessels.

Bow thrust arrangement: retractable azimuth thruster

Recent tonnage has seen azimuth thrusters (see Figure 6.20) being fitted as additional control elements to vessels already established with tunnel thrusters. This not only increases the thrust power, but places a directional force under the control of the ship handler.

Other units are fitted as retractable thrusters and engaged as a stand alone unit. Clearly a disadvantage with a thruster that projects below the line of the keel is when the vessel is navigating with reduced underkeel clearance where the thruster unit might not be able

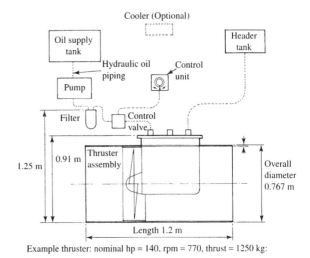

Example thruster: nominal hp = 140, rpm = 770, thrust = 1250 kg:

Figure 6.17

to be brought into operation. Also if a vessel was experiencing a minimum level of underkeel clearance and the thruster could operate, the additional problem of bottom debris causing damage to the thruster is an increased possibility.

Use of tugs in pilotage operations

Introduction

The use of tugs in pilotage has been a common practice in all the maritime nations of the world. The type of tug and the assisting methods used vary between differing ports but it is now widely recognized that the tug is another working tool of the pilot or the ship handler. Whether the tug is engaged for 'braking' the forward motion of a vessel or pushing to hold against a current will reflect the common practice of the port and its different handling methods.

It is of paramount importance that marine pilots and ship's Masters as end users of these basic tools are familiar with the working procedures and the capabilities of operations with tugs. The judicious positioning of tugs is an essential aspect towards achieving a successful manoeuvre of the parent vessel, and in itself becomes an art.

Practical use of tugs

The handling of tugs carried out from the parent vessel requires a level of mutual understanding between

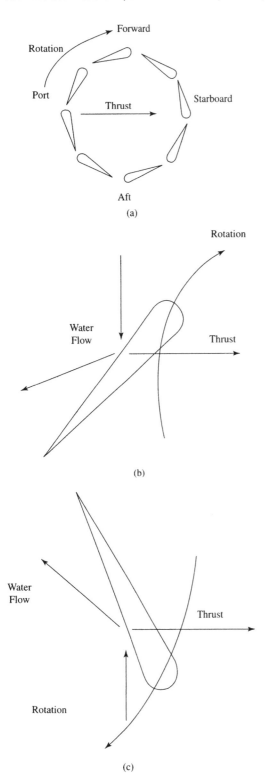

Figure 6.16   Voith Schneider operation

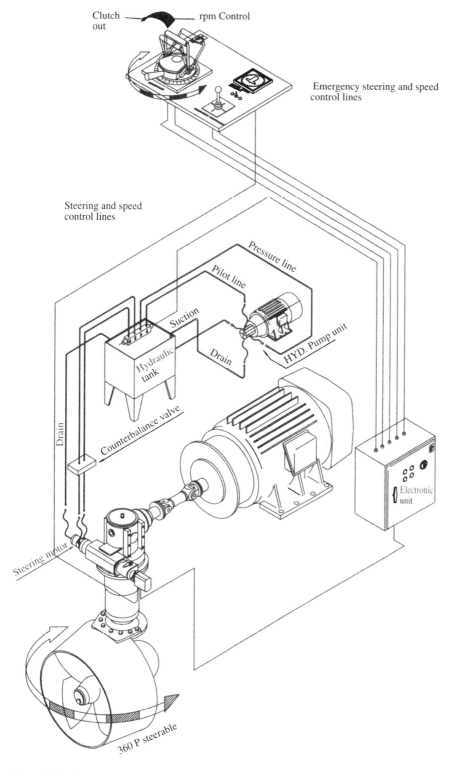

Clutch out — rpm Control

Emergency steering and speed control lines

Steering and speed control lines

Pressure line

Pilot line

Suction

Drain

Hydraulic tank

HYD. Pump unit

Drain

Counterbalance valve

Electronic unit

Steering motor

360 P steerable

Figure 6.18 Arrangement and control of 360° variable rpm steerable propulsion unit

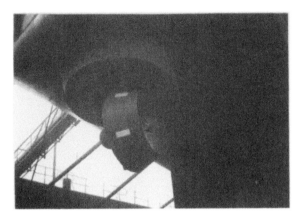

Figure 6.19   Azimuth thruster seen underside aft on a vessel in dry dock (sacrificial anodes are visible on the ducting)

the pilot/Master and the tugmaster. To this end communication of the basic needs and requirements of the mother ship becomes a vital requisite between the two stations.

Not only will the tug need the ability to manoeuvre in close quarter situations but also in restricted areas, possibly in adverse weather conditions. At the same time there will be a positive need to achieve specific objectives to move the handling operation forward.

An example of this can be seen when a tug is approaching a vessel underway, with a view to rendezvous and engagement. The tugmaster and his crew must have a clear appreciation of the instructions being passed from the parent vessel, with regard to the following:

(a) The securing method between tug and mother ship (see Figures 6.21 and 6.28).
(b) When, where and how to take a bow line.
(c) When, where and how to take a stern line.
(d) Holding station, without pulling or pushing (often alongside).
(e) The proposed speed and actual speed of vessels during engagement.
(f) Parent vessel swinging and tug response (actions to prevent 'girting' situations).
(g) Approaching the berth. Position at berth and function in prevailing weather.
(h) Differences between vessels under power compared with dead ship manoeuvres.
(i) Own tug's fendering and the vessel's structural strengthened areas, especially for pushing operations.
(j) The dangers of interaction not only with the parent vessel but also between additional tugs engaged with the same operation.

The ship handler's knowledge of the tug

The common understanding between the bridge of a mother ship and the bridge of the tug can enhance overall performance between the two vessels. The essential communication link once established can be more gainfully employed if each party to the operation is aware of the capabilities of the other and to this end the pilot or Master should be knowledgeable as to the following:

1. The type of propulsion system fitted to the tug and mode of operation which could affect response time in manoeuvres.
2. The horse power capability when pushing or towing. Practically, a Master needs to estimate what size of tug (regarding power) to use when handling a mother ship in both light and loaded conditions. Tug limitations, when engaged against strong winds, eddies or current flow, could be critical at various stages of a manoeuvre.
3. The experience of a tugmaster and his crew with respect to:
   (a) the type of vessel being controlled,
   (b) the methods of control being applied,
   (c) the geographic area where the navigation is expected,
   (d) external influences on the controlled vessel, e.g. weather.

NB. 'Bollard pull' tends to reflect tug size. It is defined as the amount of force expressed in tonnes that a tug could exert under given conditions. The static bollard pull is usually determined by a test employing a load gauge or dynamometer as part of the towline. Various factors affect the result, for example depth of water, size of tug, types of propellers, etc.

The purpose of a tug owner who quotes bollard pull is usually to provide indication of the tug's towing power under expected normal operating conditions. Depending on the speed factor the effective bollard pull could be expected to fall away.

The use of anchors on large vessels

Many VLCC/ULCC type vessels are fitted with very large, heavy anchors – up to 20 tonnes per anchor would not be considered exceptional. The associated anchor fitments are by necessity of a compatible size and weight. Cable links, for instance, are that much larger than, say, on smaller, conventional cargo ships. As with anything 'heavy' greater care, increased manpower and deliberation must be considered. Subsequently many

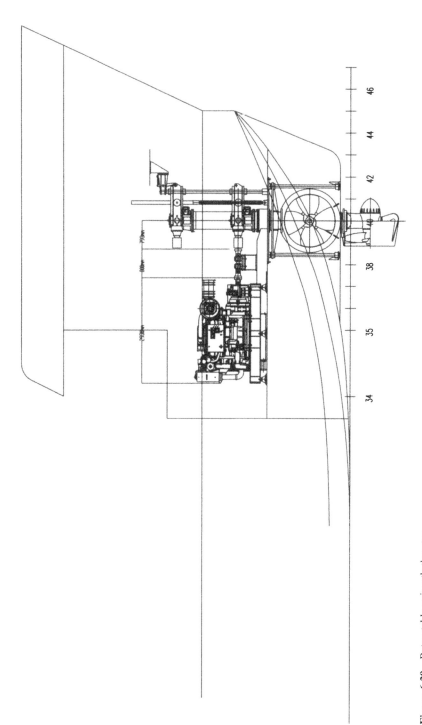

Figure 6.20   Retractable azimuth thruster

Figure 6.21 Receiving and securing the towline from a 'mother ship'

Masters of large vessels are reluctant to anchor the vessel as a first choice.

The windlass arrangement for a vessel equipped with heavy anchors must have adequate power to recover anchors and cables, and when equipment is new problems are not expected to arise. However, with older equipment the quality and capabilities could affect practical performance when both recovering and controlling anchors.

Most large vessels are expected to navigate in greater depths of water and many VLCC anchorage areas are in water depths of 40 to 60 metres. If basic principles are followed the amount of cable that should be payed out would normally equate to between six and 10 times the depth of water. The possibility that the vessel will not carry enough cable length in the locker could become a reality for some ships in certain geographic locations.

The windlass braking system must be considered a critical element of any anchoring operation. The use of 'band brakes' with smaller anchoring systems has proved successful over many years but with the new heavier operations the need for alternative brake and control systems became apparent. Disc brakes were evaluated in 1982 together with hydraulic pressure systems based on the speed being proportional to the amount of braking power applied via a control system. Although disc brakes need an increased number of 'brake pads', and larger discs need to be fitted to the heavier systems, they were generally a welcome alternative.

The need to conduct a slow walkback by ship's officers has become the normal procedure with virtually all large vessels. Maintenance of brake pads, oil seals, and the mechanical linkage of band brakes is essential for safe operations to be maintained. Some disc brake systems have experienced 'brake fade' in wet conditions and personnel carrying out anchor operations should be wary and sight the brake pads prior to walking back.

Use of anchors: associated problems

The need to use anchors should always be conducted after establishing an acceptable 'anchor plan' but even when the operation has been well thought out problems can still arise. The possibility of a 'foul hawse' (discussed in *Seamanship Techniques*, Part 1) is ever present when mooring with two anchors, and the unexpected fouled anchor on recovery is a situation which continues to occur. How these problems are eventually resolved depends greatly on the resources available and the nature of the encumbrance.

Figure 6.22 shows a cable badly fouled on itself around the head of the anchor. Clearly not a welcome sight on heaving the anchor clear of the surface. The obvious need for the term 'anchor, sighted and clear' becomes readily apparent when faced with such an unwanted obstacle.

This particular example was resolved by engaging a tug to take the fouled cable onto its after deck. The cable was broken at the joining shackle(s) and the ends heaved clear by means of messengers and the tug's winch. The kentor shackles were resecured at the ganger length and

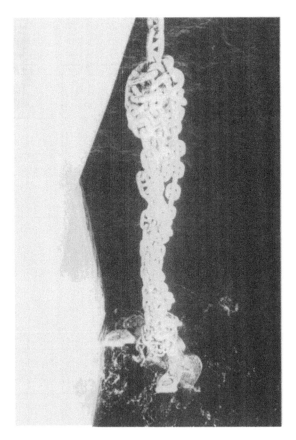

Figure 6.22 Anchor cable badly fouled around the head of the anchor

the first shackle. The cleared anchor and cable were eventually returned to the usual stowage position.

The incident caused a lengthy and unforeseen delay not to mention the additional expense of employing the tug.

This type of problem is bad enough with a medium-sized anchor, but if a similar situation was to be experienced with a larger, heavier anchor, the handling and method of clearing the foul could result in the necessity to burn through the cable.

Wheelover points

With the smaller, more manouvrable vessel the need to consider the use of 'wheelover points' prior to attaining an alteration of course position is not as great as, for example, when handling a VLCC. The larger tanker type vessels need to consider the 'way points' of the vessel's track with care and these positions where the helm is placed 'hard over' are to be incorporated in the passage plan.

A wheelover position would be established by use of the ship's data, and in particular with regard to the 'advance' and 'transfer' of the ship's turning circle. The turning circle data is established on trials and is cross-referenced against the angle of alteration required by the vessel (see Figure 6.23).

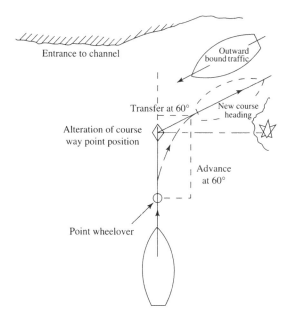

Figure 6.23 The wheelover position established using the 'advance' and 'transfer' of the ship's turning circle

The theory of establishing the wheelover point by use of the ship's data, namely the turning circle factors, can

be considered reasonably accurate provided the wheel is placed in the hard over position. (Turning circle data is compiled on trials with the wheel placed in the hard over position.) It would follow that if the wheelover position is to be established in practice then the same criteria of speed and having the wheel hard over must be applied for the data being used to establish the wheelover point for the arguments to remain valid.

In practice there are too many variables to be able to accept the theoretical wheelover position as exact. The state of loading, weather conditions, speed of vessel over ground would all be at variance with the data extracted from the ship's trials information, and therefore the theoretical position would not be based on comparing like with like.

It should also be realized that when handling a large vessel like a VLCC, a delay error would exist before the ship's head acts in response to the wheel operation. The fact that the wheel may not be placed in the hard over position for each and every alteration of course is a fact of pilotage and if placed hard over could run the risk of the ship's head being 'overcarried'.

The example in Figure 6.23 clearly identifies the need for the inbound vessel to attain the required track when approaching and entering the channel; however, due consideration must also be applied to undercutting the turn area from the theoretical wheelover position up to the time the vessel attains the new heading. Adequate underkeel clearance would be required throughout the turn and the dangers of interaction and in particular 'bank cushion effect' would also need to be avoided.

Advance planning by projecting the alteration point further to seaward would generally relieve the problems of interaction and underkeel clearance, if such sea room is available. Alternatively, a pilot would make full use of communications to ensure the ETA at the way point would not be in conflict with outward bound traffic and adjust the vessel's speed when approaching the period for the alteration.

The experienced ship handler will anticipate that different ships will respond at a rate which will reflect all the conditions on board as well as those conditions external to the vessel. Identifying at what point the wheelover position becomes appropriate and at what time the helm is eased during a turn will be acquired with practice, pertinent to the vessel engaged at the time. The same vessel, at another time and in some other state of loading, experiencing greater windage, for example, would need to be conned on the merits of the situation at that specific time.

Squat

Squat is considered a form of interaction between the sea bottom and the vessel which directly affects the vessel's draught. It is accentuated in shallow water especially in canals or rivers where the ship itself presents a blockage factor (see *Seamanship Techniques*, Part 2, Chapter 9). Different types of vessels will have a tendency to 'squat' by the head or by the stern and this will largely depend on the 'block coefficient' value.

For example, a large tanker or bulk carrier with a block coefficient value greater than 0.7 would tend to squat by the head because the longitudinal centre of buoyancy is usually forward of the centre of flotation. Whereas a passenger or container vessel with a block coefficient of less than 0.7, with the longitudinal centre of buoyancy normally found aft of the centre of flotation, would tend to squat by the stern.

NB. Where a vessel experiences reduced buoyancy effects, i.e. narrow channel, it must be expected that the longitudinal centre of buoyancy will move and as it does a trimming moment would also affect the vessel.

The amount by which a vessel can expect to increase her draught, due to experiencing the effects of squat, has been directly related to the speed squared that the vessel is operating at. Other factors, such as the type of bow, the ratio between draught and the depth of water, the ratio between the vessel's length and breadth, the blockage factor inside the channel, etc., all influence the numerical value of reduced underkeel clearance that must be anticipated. However, the ship's speed is the main influential factor and it is a factor that the ship handler can do something about.

When a larger vessel is considered in conjunction with 'wheelover position' there is a conflict of interest in the fact that underkeel clearance could be reduced if the change of heading at a 'way point' is not carefully planned in advance. The movement would need a degree of speed to maintain steerage effect in the turn and this could in itself incur an additional increase in draught because of the effects of squat. If in a narrow waterway, heeling effects coupled with the effects of squat in a tight turning situation could lead to the ship touching the bottom in the area of the turn of the bilge.

Manoeuvring large vessels

The larger vessel has become the accepted norm within the marine environment and the heavy bulk carriers and long supertankers have common place in the world's shipping lanes. Deep water routes towards ports tend to segregate the more general shipping from the very large vessels but this is not universal and it must be expected that the coaster and the VLCC will work in close proximity on occasions. It is therefore beneficial

for anyone in a bridge team to become aware of each other's problems especially with regard to manoeuvring either in open or confined waters.

The very length of the larger vessel can place the bridge Watch Officer at a distance in excess of 300 metres from the bow area. Such remoteness especially inside the confines of an enclosed bridge could lead to a false sense of security with heavy seas breaking over the bow. Experience of the handling characteristics of the large vessel can be acquired with practice, but solid water breaking over a foc's'tle at night could remain undetected resulting in subsequent damage.

All vessels experience stress forces when in heavy seas. Depending on the ship's length overall, the direction of sea and swell waves will determine the type and value of stresses that the ship is being subjected to. Rolling motions from beam seas are likely to cause 'racking stresses' athwartships, while bow seas could give rise to 'torsion' stresses through the vessel's length. The additional stresses brought on by 'hogging and sagging' together with bending and shear forces could be accentuated by a poor loading condition and Masters are advised to keep a watchful eye on the loading programme being implemented from the onset of loading.

Heavy weather

If encountering 'heavy weather' the large vessel could expect to experience high impact forces in the bow area. Coupled with a deep draught the pressures generated with an excessive pitching motion could be considerable. The speed of the vessel over the ground will be seen to be reduced from impact forces and the Watch Officer should subsequently reduce the load on the engine and additionally reduce engine speed. Where the direction of the sea is from right ahead a course change to place the wind two to three points on either bow might lead to a more comfortable heading and reduce stress levels.

A reduction of speed is probably the most influential action and such a reduction could be of such a magnitude as to just retain steerage way on the ship. The weather would need to be continually monitored and speed adjusted to suit circumstances.

Turning the large vessel

When a VLCC or other similar sized vessel is turned, the performance in the turn will vary depending on several conditions, e.g. deep water or shallow water, speed of the vessel when entering the turn, rudder response time, position of the pivot point, tidal or current conditions prevalent, the draught of the vessel and whether she is loaded or in a light condition.

Turning circles achieved for ship's trials are ideally conducted in good weather conditions with little or no wind, with the vessel in a light condition. Whereas once the vessel is operational, it will probably be loaded and weather conditions at the time of the turn cannot be guaranteed and will probably be far from ideal.

When the ship moves forward and increases speed, a level of water resistance will be built up ahead of the vessel which causes a balance to be established with the vessel's power. This build-up of forward resistance and resulting 'balance' causes the pivot point of the vessel to be set back by up to an additional quarter length of the vessel. This effectively places the position of the pivot point of the vessel at about one-third of the vessel's length from forward and reduces the turning lever of the vessel. This results in the force of the rudder becoming less efficient and the vessel experiences a sideways motion in the turning movement. Such a sideways motion is met with increased lateral resistance forces (see Figure 6.24).

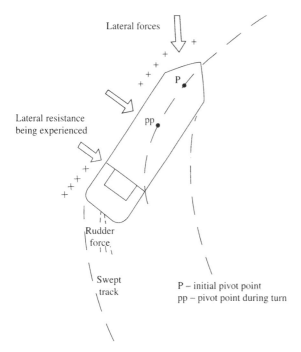

Figure 6.24 Turning the vessel

In any turn, the efficiency of the rudder will be checked by encountering higher lateral resistance forces, so by entering a turn at a higher speed, the higher will be the lateral forces experienced. The only real change will be in the time element, whereas the advance and transfer of the vessel in the turn will generally stay about

the same. However, the 'rate of turn' will vary and this could be critical if turning across a strong tide and as such may not improve the overall turning ability of the vessel.

Speed of turn

The majority of vessels expect to experience a speed reduction if altering the ship's head to a helm hard over position. As the ship moves in a sideways motion the hull side meets increased water resistance as compared with the bow. This effectively acts as a brake and could reduce the vessel's speed by as much as 50 per cent. This effect can and is gainfully employed when it is a requirement to reduce the vessel's speed over the ground. Pilots sometimes use two 90° turns to bring a vessel's speed down, as when approaching an SBM, and the reduction would also be desirable when executing a man overboard, Williamson turn.

Such manoeuvres could only be considered if the sea room is available. Where sea room is known to be restricted prudent early reduction of main propulsion is advised as a means of attaining the desired speed of approach.

Certain vessels such as high decked ferries or container vessels could be influenced by wind direction when conducting a turn. This could create or extenuate any list on the vessel during the turning operation and cause possible draught or interaction forces to be brought into the equation, especially with shallow water turns. Well-loaded, low lying tankers tend not to have the same windage problems that may be experienced by the intermediate-sized vessel but the deep draught of such vessels, if also influenced by squat, might result in the vessel having to pull back on speed to avoid compromising its underkeel clearance during a turn.

Figure 6.25 A large tanker vessel in open waters, the effect of water resistance can be clearly seen across the vessel's bow

(a)

(b)

Figure 6.26 VLCC: approaching an oil floating storage unit (FSU). (a) Pipe boom, supported and extended from the FSU. (b) VLCC on final approach to FSU, prior to picking up a light messenger in order to pass heavier moorings

Figure 6.28 Large tanker manoeuvring with four tugs assisting

Figure 6.27 ULCC Shell oil tanker laying to a single buoy mooring (SBM) in a light condition. Loading pipes are visible overside, just forward of the midships position

Safe speed conditions:

When considering what constitutes a 'safe speed' all vessels should make reference to the content of Rule 6 of the anti-collision regulations but in general large vessels would by necessity also take account of:

1. The vessel's manoeuvring capabilities, and in particular the stopping distances relevant to various speeds and respective response times.
2. The current state of visibility.
3. The traffic density.
4. The available depth of water required to maintain an adequate underkeel clearance.
5. The available sea room and width of any traffic lane or channel.
6. The level of reliability of steering and main machinery equipment of the vessel.
7. The proximity of additional navigation hazards (e.g. fishing boats or leisure craft).
8. The need or intended use of anchors (large vessels are equipped with exceptionally large anchors often

in excess of 20 tonnes and if employed would not usually be 'let go' but walked back in gear).

9. The direction and rate of any tide or current flow acting on the hull.
10. The direction and strength of wind, particularly so when the vessel is in a light or ballast condition.

Where traffic conditions are expected to be more concentrated, as when making a landfall, it should be noted that the combined speed of an oncoming target operating at the same speed as a large vessel will cause a potential impact position to be reached in less than half the stopping time of own vessel. This is assuming that the target vessel is making headway at least equal to the large ship's speed. Neither will a reduction of speed over the ground occur at a uniform rate, so the time of actual contact could be expected to be even less on the assumption that a vessel reduces her forward motion more slowly at the termination of headreach than at that moment when first starting headreach.

Navigation in poor visibility

The associated dangers of navigation in poor visibility are well known to every mariner. Proceeding in 'fog' always has an element of risk but proceeding at excessive speed in a large vessel, knowing the stopping distance and the time involved, must be considered as unwise.

Because 'crash stop' action takes some distance to achieve, the alteration of course is generally more widely used by a conventional vessel. The large VLCC which detects a close quarter situation developing may wish to make a course alteration (knowing the advance in a turn will probably be less than the distance it would take to stop); however, the vessel may not be able to exercise this option because of a deep draught and/or restricted sea room. Also, too small an alteration may not be readily apparent to another radar observing vessel, yet a small alteration, because of the proximity of shoals, may be all that the large vessel can afford to execute.

Clearly the benefits of VTS control operations and effective communications can be employed in conducting a safe passage in any state of visibility but the dangers of acting on radio instructions should not be condoned without positive station identification and/or the confirmed position of a transmitting vessel. Extreme caution is advised with VHF in good visibility, and its use for collision avoidance in poor visibility must be considered questionable unless station identification can be guaranteed.

Many Masters obviously prefer to keep their starboard sides clear but this is not always a viable option, especially with the larger type vessel navigating with draught restrictions. Advance planning, to prevent vessels meeting in narrows, and early adjustment of speed must be recognized as an act of good seamanship. Masters doing 'too little too late' especially when faced with restricted visibility may regret not reducing speed sooner to suit the changing conditions.

One consolation for the ULCC Master is that the more conventional-sized vessel is not looking for a close quarters situation to occur, any more than a large vessel is anxious to close with another target vessel. Deep water routes provide some welcome segregation but the leisure yacht, without a radar reflector, and with a fibre glass hull is not the easiest of targets to avoid even in moderate seas.

The disabled large tanker

The loss of power or steering off the 'lee shore' is a time that the Master earns every cent of his salary. There is no right solution but a few suggestions are offered towards a possible procedure which might relieve the worst scenario of a total constructive loss.

Once advised of the nature of the disability affecting the ship a prudent Master would take the 'con' of the vessel and order a position to be placed on the chart. NUC signals would be displayed and either the Chief Officer and/or the Chief Engineer would be instructed to make a damage assessment report.

An updated weather report should be obtained and the current situation assessed with regard to the distance off navigational hazards. These circumstances could well dictate the type of communications that may be possible.

The reader will quickly realize that such actions would probably be carried out with every vessel faced with such a predicament. However, the degree of 'urgency' and the speed of activity by the Master of a large tanker cannot be overemphasized.

Once a damage assessment has been made it would be prudent to instigate repairs as soon as possible, if this is practical. Clearly, certain circumstances might make this an impractical option as when the ship's rudder is lost or an engine room explosion has left main machinery inoperative and beyond repair. The smaller vessel might consider some form of jury steering gear to replace a lost rudder but the large tanker would have to obtain ocean tug assistance.

Deck operation of preparing the emergency towing equipment should be ordered (tankers of 20 000 tonnes or more must now be fitted, fore and aft, with a strong-point and chafing chain to allow tugs to secure to – see Chapter 5, 'Smit brackets'). Special signals, navigation lights and shapes should also be prepared prior to commencing any towing operation.

The nature of the incident will depend on the choice of port of refuge. Dry dock facilities are usually an essential requirement but not every port will have dry dock capability for such a large vessel. With this in mind, consultation with owners/agents in establishing a suitable destination would need to be a priority before the tug's arrival on site. The Master would further need to re-route and prepare a revised 'passage plan' to be compatible with the vessel's draught and size, and take account of any degree of damage which might affect the ship's navigation under tow.

The Master would be faced with preventive holding action once tugs are confirmed to be on route. Such holding activities could include one or several of the following options:

1. Ballast the fore end to weather vane the vessel (sometimes called 'flag effect'). Large vessels have large tanks and it may take some time to ballast the fore end with the view to increasing the windage aft. If propellers are still operational, headway could be made to increase the distance off a shoreline.
2. Walk back both anchors to below the keel depth to provide a drag effect in the bow region allowing the wind to act on the aft accommodation block. If the vessel heads the wind use ahead power if available to maintain distance off.
3. Depending on depth, attempt to establish a 'deep water' anchoring position with one or both anchors.
4. Use a small vessel (probably inadequate to establish a tow), for steerage or standby. A well-fended bow might be able to push the tankers bow upwind. Alternatively, a small tug made fast astern, which may not have sufficient power to tow the larger vessel, could possibly provide steerage towards safer waters if the tanker has retained main engine power.
5. If main engine power is still available, the option to 'stern bore' the vessel into the wind is a further possibility, provided adequate sea room and geography permit such a manoeuvre. Some damage to the aft end of the vessel might occur if heavy seas are present at the time; however, limited damage might be acceptable if the astern motion can keep the vessel away from the shoreline.
6. Where a large vessel has lost all power, and is drifting without possible chance of repair, the danger should be recognized swiftly and tug assistance summoned sooner rather than later.

Each incident would of course need to be judged on its merits and would depend on what remaining capability the ship has, what equipment is available and functional and the geography of the situation. The time interval will be determined by the weather conditions, current/tidal rates affecting the vessel and speed of practical response by the Master or officer in charge.

Communications for the disabled vessel

The nature of the incident and general circumstances will all influence the type of communications made both internally and externally from the ship. Keeping persons on board informed of the ongoing situation can avoid speculation and unnecessary panic. It is also appreciated by crew members that, more than ever, mutual co-operation could help save the day.

Visual NUC signals or use of international code flags keep local shipping with line of sight, informed and make the vessel more easily recognizable when tugs or other support/rescue vessels are approaching. These signals can be displayed reasonably quickly even while a damage assessment is ongoing.

External communications by way of an 'urgency' or even a 'distress' message should not be envisaged until the ship's position is ascertained and the Master is in possession of the damage report. Once contact is established there are bound to be questions and possible clarification of facts from the shoreside authorities for which the Master needs to know the answers.

Masters normally keep owners and agents informed through coast radio stations and instruct reports to be passed to underwriters, P & I Clubs, Charterers, shippers, MAIB/MPCU, and other affected parties. On a more practical aspect communication with a tug operator to arrange either a contract of tow and/or salvage request would be a priority. (Owners may well assist with towage requests as would most Coastguard administrations.)

The Port of Refuge, dry dock authorities, pilots, VTS operators, or other ship reporting systems (AMVER, AUSREP, etc.) should all be appraised of the ship's status and any deviation being experienced. Revised routeing information, once the immediate danger is resolved, should be passed to interested parties. Weather reports for the immediate area together with facsimile and prognostic isobaric charts should also be obtained if possible.

Whether a situation warrants an 'urgency' or a 'distress' signal is judged by the Master. However, it should be realized that an 'urgency' signal can always be upgraded to a 'distress' ('priority 1') signal if the situation deteriorates beyond the scope of an urgency message. The rate of drift towards a coastline and the type of vessel involved clearly influences the priority of the call, as would the distance off a navigation hazard. Large tankers and passenger vessels would in virtually every case go to an immediate 'distress' in order to provide as much time as possible to reduce the risk to life and damage to the environment.

High speed craft

Following developments in high speed vessels, IMO introduced the High Speed Craft (HSC) code in 1996. The main concerns were initially focused around catamarans or semi-displacement mono-hull craft carrying passengers in restricted waters. However, developments of larger craft with increased passenger payload have allowed such vessels into greater and more unrestricted waters (see Figure 6.29).

Figure 6.29 The *Stena Explorer* high speed vehicle/passenger ferry engaged on the Irish Sea trade, navigating stern first towards the shaped and fended berth at Holyhead

Such developments generated considerable interest in the durability of the hull, because of the lighter construction of the different hull forms involved. Many designs use aluminium features, and leave large areas unpainted to reduce overall weight. These hulls have high values of reserve buoyancy and a limited number of openings that could promote downward flooding in the event of an accident all of which has led to higher safety levels.

To date, no high speed craft has sunk or capsized following collision or grounding where watertight integrity of the hull was lost. However, acute listing has been seen to develop, especially on catamaran type hulls, and this unacceptable angle of heel could prove a problem when evacuating large numbers of passengers. More thought is therefore needed to keep an even keel on vessels which in theory would seem to have an acceptable level of flotability.

The high speed of such craft is clearly a commercial asset, but the prime function of the HSC code must be the overall safety of the passengers and crew of such vessels. Designs are moving towards keeping passengers and emergency equipment away from areas of the hull susceptible to damage from either collision or grounding. Another concern is the protection of passengers

from sudden deceleration effects, e.g. hovercraft moving at about 80 knots.

The more obvious dangers in the marine environment from fire or evacuation, creating the need to abandon, are also catered for by way of each passenger accommodation space having at least two unobstructed access points. Fast ferry type craft tend to have open public spaces where any fire outbreak is easily detected and generates an alarm, although machinery spaces remain as vulnerable as any other conventional vessel.

The inherent high speed of such craft provides a means of avoiding close quarters situations by manoeuvring out of range in open sea conditions, but in restricted waters where speeds are reduced the 'outrun' option is not always possible. When in close proximity to other ships the noise from certain craft, hydrofoils, jet boats or hovercraft, may reduce the effect of sound signals. Such craft may also have difficulty in displaying navigation

Figure 6.30 Catamaran with centre line anchor position, engaged on the Irish Sea ferry trade

Figure 6.31 Centre line securing bracket at the stern of the *Stena Explorer* for reception of a quayside holding bolt, effectively clamping the vessel into exactly the same position each time she berths. This ensures that the vehicle ramp/rail lines, etc. are aligned to perfection prior to commencing the loading/discharging of vehicles

lights in the more generally accepted positions and as such may need to shelter behind Regulations 1 and 2 of the anti-collision regulations.

Mooring operations

Although many high speed craft have conventional equipment for completing mooring operations by way of the usual anchor arrangements, mooring ropes and tension winches, other systems have been designed to suit the trade.

Some vessels where rail traffic is shunted and carried aboard necessitate that when stern to, and mooring up takes place, the exact position of the ship is required to ensure rail track lines ashore and on board are meeting true. Such exactness has been achieved by shaped docks and fixed stern securing, which in practice clamps the ship's stern into a positive position by means of locking bolts (see Figure 6.31). The vehicle ramp is then lowered into the horizontal position to allow access for roll on-roll off traffic into the vehicle deck levels, passenger accommodation being usually set at high deck levels well above vehicle decks.

7 Engineering knowledge for Masters

Introduction

It is not the intention of this chapter to attempt to turn the Master into a Chief Engineer. It is merely meant to familiarize the seagoing Master with some basic elements of the activities and developments in the subject of marine engineering. Most serving Masters would have at some time in the past experienced the request from the engine room to reduce speed because of a 'scavange fire', but the question should then arise, what caused that fire? The fact that the most common cause is worn piston rings begs the question, why is the vessel operating with worn piston rings in the first place and is the function of planned maintenance fulfilling the vessel's needs?

The Master should be equipped with a basic understanding of marine engineering and fundamental control systems within his vessel; however, he should not lose the essential assistance and loyalty of his Chief. 'Ship command' has been defined as the authority to direct and control a ship, the purpose of which is to carry out the management of all shipboard operations, not least the indirect operation of the vessel's machinery.

It is generally when things don't always go according to plan, for one reason or another, that the ship's Master and Chief Engineer tend to come together. The problems of faulty steering or loss of sea speed are typical examples when the Master could expect feedback. An understanding of terminology, reference to components and the ability to acknowledge one's own limitations could well go a long way towards identifying the problem and instigating corrective action.

Main engine power

In order for the Master to function as a ship handler it is essential that main engine power remains available. It is appreciated that a vessel without power can be controlled by external forces, supplied by tugs, but under normal circumstances a Master expects to have full use of main engines. An understanding of the power that is being produced by the engine and the relationship to the actual propeller power being achieved, to provide the ship's motion, differ considerably.

Definitions

'Indicated power' is defined as the actual power developed in the cylinders, derived from the high pressure, hot gases acting on the piston area.

Indicated power is usually measured daily by the engineering officers to ensure satisfactory performance of all cylinders. To obtain the indicated power of an engine cylinder an 'engine indicator' is fitted to a test cock on the side of the cylinder. The device produces an indicator diagram and the measured area represents the pressure acting on the piston through its stroke.

'Shaft power' is defined as that power available at the engine output shaft, used to drive the propeller or machinery.

NB. Shaft power will always be less than indicated power because of friction and heat losses that occur within the engine.

Shaft power of an engine can be obtained by employing a 'torsion meter' which is used to measure the turning force (twisting force) acting on the shaft. The obtained value is then used to calculate the shaft power being achieved.

'Propeller power' is defined as the actual power developed by the propeller due to the revolutions and the pitch angle of the propeller blades.

NB. Propeller power will always be less than the 'shaft power' because of friction from bearings and stern tube construction and 'blade slip' in the water.

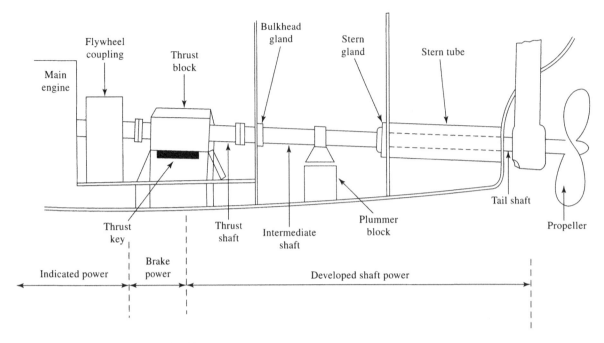

Figure 7.1   Main engine to propeller drive

Figure 7.1 shows diagrammatically the components needed to produce motion at the propeller from the main engine, and Figure 7.2(a) and (b) show typical engine control room console displays.

Engine room records

In the event that a Master is called upon to give evidence to an inquiry or court hearing all sources of information may be required for inspection. These sources may include contemporaneous records such as rough log books and movement books. In the case of the engine room such records may be in the form of:

Data logger information sheets.
Engine room telegraph printouts.
Bunker records.
Engine Room Log Book.
Movement book.

NB. Vessels with less than 150 kW registered power or vessels with unidirectional main propulsion driving controllable pitch propellers whilst on bridge control, and vessels with engines operated by remote control units active from the bridge, are not required to maintain a movement book.

Where vessels are under bridge control, records for engine movements should be contained alongside course

(a)

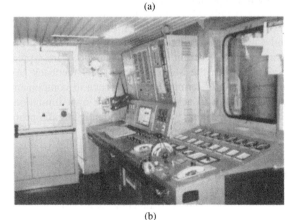

(b)

Figure 7.2   Engine control room console display types

records, echo sounder graphs, telegraph printouts, NAV-TEX reports, etc.

Plant monitoring: unmanned machinery space

With so many different designs of bridge control now in operation it would be unrealistic to lean towards a single system, especially as most operations incorporate all the basic elements needed to provide dual position control. The fundamental system will include bridge instrumentation which monitors and controls:

(a) Propeller speed.
(b) Propeller direction of rotation and/or pitch angle of CPP blades.
(c) Audible and visual alarms on the bridge and in the engineering environment to signal power failure, or other essential control element. (These alarms may be relayed via engineer's accommodation.)
(d) In the case of a diesel engine, 'air start pressure' sufficient for manoeuvring.
(e) An independent means of stopping propulsion systems in the event of an emergency.

The alarm system

A total alarm system alerts both engineers and the bridge (when the navigator is the sole watchkeeper) when:

(a) a machinery fault occurs,
(b) the machinery fault is being attended to,
(c) the machinery fault has been rectified.

The monitoring of systems such as power failure and similar elements will have audible and visual alarm coverage, with means of silencing the audible alarm, but allowing the visual alarm to remain until the fault is cleared. In the event that a second fault develops while the first is being dealt with, the audible alarm is reactivated.

An alarm indicator at a remote station from the machinery space, i.e. bridge, will not automatically silence the same alarm inside the machinery space – each and every alarm display once activated will identify the particular fault at either the main control station or at a secondary station. This effectively increases the more positive response in a shorter time period.

Specific alarm coverage

Where high risk areas such as those subject to fire or flooding are monitored, it would be a requirement to have two independent monitoring systems interfaced with communications, e.g. bilge level alarm system with direct communication to the duty engineer's cabin, or a fire alarm from the machinery space direct to the bridge watchkeeping station.

Fire/heat detection systems

Figure 7.3 clearly shows a flame sensor/detector device, in this case fitted to a vessel's engine room.

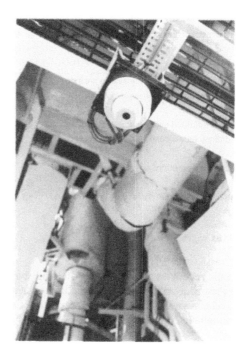

Figure 7.3 Flame sensor/detector fitted to ship's engine room. (Manufactured by AFA Minerva Ltd, Marine and Offshore Division)

Mode of control

It is generally recognized that for a vessel fitted to UMS standards, the normal mode of control will be from the bridge. To effect a change of control from the bridge to the engine room (see Figure 7.4) or vice versa is normally achieved from the machinery space itself. Control station status is of course only initiated with direct communication and on the understanding that the respective station is ready to accept that control.

In the event of system breakdown, the engine room has the capability to revert to full manual control with human intervention. If such an event did occur Masters should be aware that the parameters that were previously automatically recorded and monitored (usually by a data logger) would need to be observed by engineering

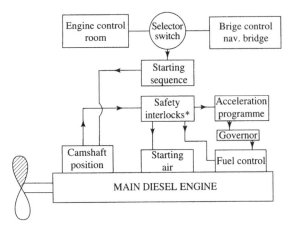

* Safety interlocks

1. No start of main engine with turning gear in position.
2. No start of main engine unless the propeller pitch is set at zero (CPP).
3. No starting air is admitted once the engine is running.
4. No fuel is admitted unless correct starting sequence is followed.
5. No astern movement permitted unless engine is first stopped.
6. Main engine cut out if the governor limits are exceeded.

Figure 7.4 Main engine–bridge control: direct reversing diesel engine

Figure 7.5 Standby emergency diesel generator turbo-charged air cooler at left hand top end (six cylinder turbo charged engine)

staff. This could reflect on response time affecting ship movement as any required change by actuators to attain the desired value would require manual operation as opposed to automatic operation.

Control station change

When changing from one control station to another it is expected that the change takes place simply and effectively. Normal practice dictates that engine room and bridge are both on 'standby' and a selector switch is turned to establish the mode of control (see Figures 7.5 and 7.6 for typical engine room features). Once on-line control is accepted the alarm system should be tested and essential control elements checked out as operational.

NB. A basic requirement of operation is that essential machinery could still be controlled manually in the event of bridge control or auto control malfunction.

Emergency fuel shut-off

Most Deck Officers are aware that the ship's main plant will have remote emergency shut-off controls in such

Figure 7.6 Top view of six cylinder, medium speed diesel engine

a position as to be readily accessible in the event of an emergency. The actual positions of these controls will vary from ship to ship but would not be out of place in the Chief Engineer's office, on an exposed boat deck or in a damage/emergency control room. In virtually every case they are a 'gate valve' operated mechanically or more usually by electronic solenoid switches, as illustrated in Figure 7.7, to cut fuel supply and so bring machinery to a stop.

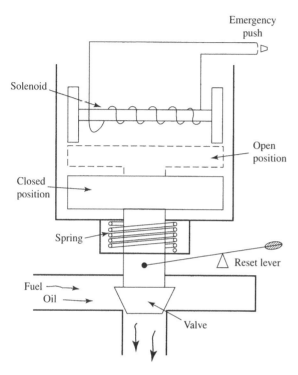

Figure 7.7 Mechanically operated 'gate valve' used to cut fuel supply

Steam turbines

Data logger links and alarm safeguards for steam turbine plant

- Lubricating oil pressure for turbines and gearing low.
- Lubricating oil temperature for turbines and gearing high.
- Bearing temperature for turbines and gearing high.
- Astern turbine temperature too high.
- Steam pressure at the gland too high or too low.
- Excessive turbine vibration, generates alarm.
- Axial movement of turbine rotor, if excessive, generates alarm.
- Sea water pressure or flow, if reduced, generates alarm.
- A low condenser vacuum, activates the alarm.
- Condensate level too high, generates the alarm.

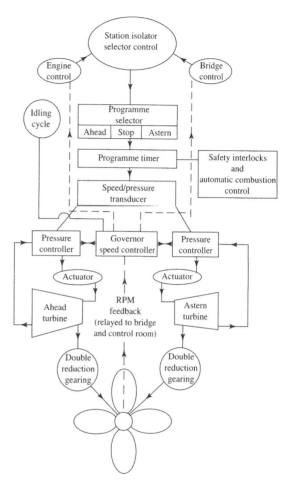

Figure 7.8 Steam turbine control system

Figure 7.9 Turbine rotor, exposed and viewed with turbine casing removed

Automatic 'stop' occurs if the lubricating oil is lost completely. Speed reduction occurs if the machinery experienced excessive vibration or axial distortion or if condenser problems existed.

Turbine operations incorporate automatic back-up systems if:

(a) there is any loss of pressure of lubricating oil,
(b) the condensate pump slows,
(c) the main sea water circulating pump slows.

The closed feed system

The principle of the closed feed water system (see Figure 7.10) is to generate steam from the water in the boiler to carry the heat energy to drive the plant. The ideal system would be to recover the steam at the end of its cycle and return it as feed water, without incurring any loss.

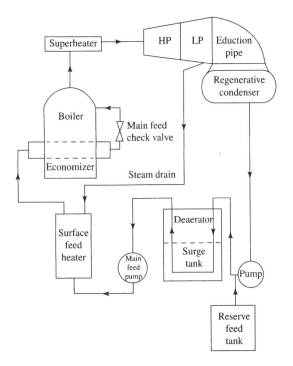

Figure 7.10   Closed boiler feed water system for use with turbines

In the closed feed system there is a need to minimize the loss of water and heat – essential savings at any time, but never more so than with a shipboard system. Some heat can be utilized to increase the overall efficiency of the plant, and to vent steam off to atmosphere would create a need for considerably greater quantities of 'feed water' to be available.

The essential aspects of the feed system are:

(a) To transfer the condensed steam (condensate) from the condenser back to the boiler via the deaerator and feed heater.

(b) To increase the efficiency of the plant by use of feed heaters which employ live steam direct from the turbine or exhaust steam from auxiliaries to increase the feed water temperature.
(c) To provide quality controlled feed water by monitored analysis in order to protect the boiler from impurities and the effects of scale.

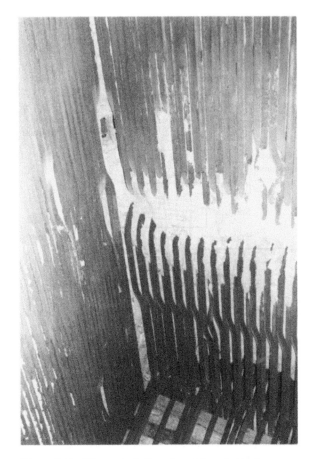

Figure 7.11   Water tube boiler, viewed from inside the furnace area

The water tube boiler, as seen from inside the furnace area, is shown in Figure 7.11.

Engine room operations and maintenance

Cascade system

The cascade system employs a Master controller, which emits an error signal (E), to activate another slave control section. A similar system monitors the level and supply of water delivery to the boiler when steam demand is high.

Figure 7.12 Piston removal from a slow speed two-stroke crosshead engine

Figure 7.13 Condenser opened for maintenance

Figure 7.14 shows boiler fuel and air delivery being generated from the pressure and flow sensors monitoring the main steam line. When demand by consumables is high an initial error signal activates the master steam controller which subsequently activates the two slave systems controlling the air and fuel.

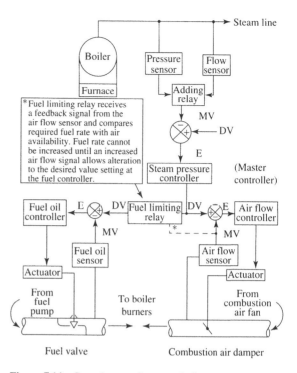

Figure 7.14 Cascade control system: boiler steam

NB. The measured value (MV) is compared with the desired value (DV). If these two values are not similar then an error signal (E) would be generated to activate the controlled element.

Boiler terms

Priming
A fault in the boiler due to overheating, where water is actually carried over with the steam. A highly undesirable condition which could lead to erosion and the cause of 'water hammer'.

Shrinkage
An apparent loss of the water level in the boiler due to a reduction in steam demand, e.g. if a vessel is steaming at full speed and drawing full steam supply, and is then suddenly reduced to half speed, a back pressure is momentarily experienced. This is detected by the water level sensor gauge which actuates a top-up by the feed water control.

Swell
The opposite effect to shrinkage where an apparent increase in the water level is caused by increased steam demand.

Engineering duties

Prior to arrival standby

When arriving at a port it is the duty officer's responsibility to keep the engineering department fully informed of the vessel's progress to ensure that when the engines are required the essential tasks for manoeuvring have been completed. Such duties vary depending on the type of engine but the following example is provided for a 'slow speed diesel'.

Normal practice includes checking the air start system, i.e. all bottles are full and ready for operation and all valves are lubricated. Prior to standby, the heavy fuel system is changed to diesel oil and the revolutions are reduced slowly to avoid damage to the pistons.

The manoeuvring handles are oiled and additional personnel placed on standby duty. Once the vessel is placed on standby, the auxiliary scavenge air pump is started and the air start system valves opened up to make the system ready for immediate use.

At full away

Once a vessel has cleared the standby area the Master is expected to order 'full away' and the engine room personnel should be actively engaged in bringing the vessel up to full sea speed.

Initially the auxiliary scavenge air pump is shut down, and the engine's speed gradually increased from full manoeuvring. The change from light diesel fuel to heavy fuel oil for normal running is carried out slowly before the vessel attains 'full sea speed'. Sea watches are set and additional standby personnel stood down.

At finished with engines

When a position is reached that 'finish with engines' (FWE) is ordered the main machinery will gradually be shut down. Initial actions entail opening the drain valves to release the fuel pressures and closing the outlet valves to the high pressure fuel filters, and closing down all fuel pump suction valves leading from gravity tanks to prevent the risk of leakage through the pump and filter drains.

NB. The head of a gravity tank could be sufficient to lift suction valves of a fuel pump.

The circulation of cylinder jackets and pistons is continued for about half an hour after stopping, to allow uniform cooling to take place. The sea water cooler and the oil lubrication systems are then stopped and shut down.

Control and monitoring of machinery and plant systems

The modern vessel is fitted with many operational systems to allow the ship to be worked as a floating transport. Cargo, ballast and fuel systems are common to all types of commercial vessel and they are all often fitted with a variety of actuators, sensors and controllers in order to function (see Figure 7.15). When the instrumentation to a plant or specific system is extensive, the task of monitoring and assimilating information becomes a major task which can be labour intensive and time consuming.

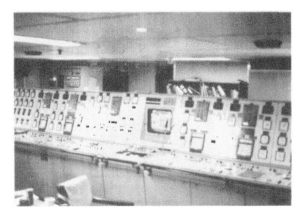

Figure 7.15   Cargo control room console display

With smaller crews and larger vessels being the order of the day the need for effective control systems becomes essential, and was positively addressed by the introduction of 'data loggers' and 'mimic diagrams'. These relieved labour problems and centralized shipboard monitoring operations to specific control rooms.

Mimic diagrams

A mimic diagram is a pictorial representation of the nature and geographical layout of a plant. Shipboard examples are a cargo pipeline system, or a water ballast line operation.

Figure 7.16 shows:

(a) One or more systems within the layout, represented by alternative colour schemes to separate functions, direction, commodities, etc.
(b) Status lights for indicating open or closed valve positions, tank levels high or low, active or non-active lines, etc.

(c) Additionally, display indicators providing readouts for temperature or pressure or other similar elements may be included with alarm acknowledgements and/or pushbutton stop/start controllers, i.e. for pump operations.

Figure 7.16 Mimic diagram – tanker ballast display

Data logger systems

Data logger systems have become a well-established feature of shipboard operations and are used to monitor all aspects of machinery plant and cargo operations. They have proved themselves ideal for specific cargoes, such as product carriers and refrigerated cargoes where the facility can effectively record cargo space temperatures.

A data logger is essentially an information recording device which allows continuous scanning of virtually any number of variables. Marine parameters which are under constant monitoring include such items as temperatures, flow rates, pressures, liquid levels, power output, torsion stresses, rotation speeds, etc. A modern system could be expected to scan up to 5000 elements per second and provide an instantaneous response by way of alarm units interlinked with the operation.

The need for this type of monitoring became essential to the development of unattended machinery spaces (UMS). The basic equipment not only incorporates audible and visual alarms but can also deliver a time identified recorded printout as standard. Levels of accuracy of recording and monitoring are improved and the possibility of human error is reduced considerably. High reliability alarm scanning can relieve potentially dangerous scenarios before they are allowed to escalate into an emergency situation.

Advantages of operating with such equipment provide:

- early warning of malfunction,
- increased efficiency by relieving watchkeepers of routine tasks,

- determination of trends on machinery,
- accurate records of operational data,
- reduced instrumentation,
- preset sensitivity and tolerance margins,
- continuous centralized observation.

Single point examination permits itemized inspection of any variable at any time by duty personnel, readouts being in analogue or digital format to suit the observer and/or the units of measure.

Equipment is comprised of four main sections:

1. measurement of variables,
2. signal and scanning communication,
3. signal processing and interpretation,
4. units of output.

Figure 7.17 illustrates the basic principle of operation.

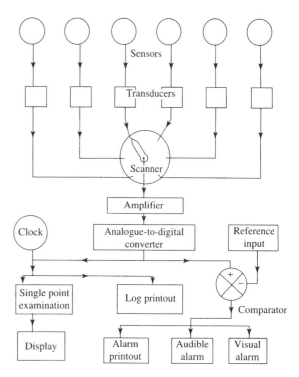

Figure 7.17 Data logger principle

On-board sewage systems

Although at the time of writing Annex IV to the MARPOL convention is not in force, several maritime nations have put in place domestic legislation which controls the discharge of sewage in the proximity of coastal waters.

Criteria for discharge

Ships are in general prohibited to discharge sewage unless it is comminuted or disinfected using an approved system at a distance of more than 4 nautical miles from the nearest land or in the case of non-comminuted or non-disinfected sewage at a distance of more than 12 miles from the nearest land.

The proviso being that sewage which has been retained in a holding tank shall not be discharged instantaneously but at a moderate rate while the ship is on route and proceeding at not less than 4 knots.

Three options exist for the discharge of sewerage:

1. The administration will approve the rate of discharge.
2. The ship has an approved sewage treatment plant in operation, the test results of which will be identified in the ship's International Sewage Pollution Prevention Certificate (ISPP). In addition the effluent must not produce visible, floating solids or cause the surrounding water to be discoloured.
3. The vessel is under the jurisdiction of a state where she is allowed to discharge under less stringent requirements.

Figures 7.18 to 7.22 show sewage treatment plants of the type that would satisfy the term 'approved system'.

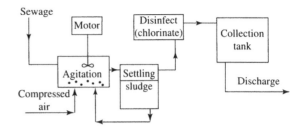

Figure 7.19   Biological sewage system

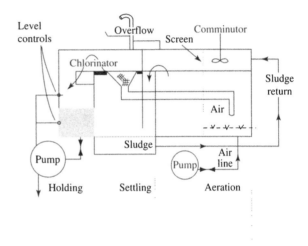

Figure 7.20   Sewage biological plant operation

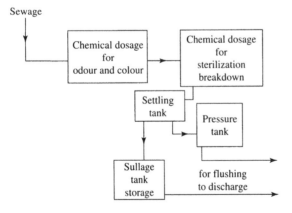

Figure 7.18   Chemical sewage, recirculation system

Oil/water separator

In Figure 7.21 the limit switches control the 'feed pump' which transfers oily bilge water to the separator inlet. The extraction pump discharges the treated water over the side, while separated oil is drawn off to a holding tank. The oil/water mixtures of the filter and sludge back flush are returned to the bilges for reprocessing.

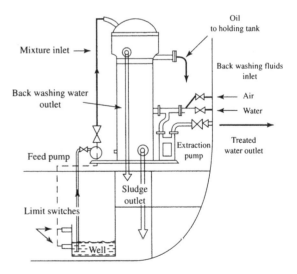

Figure 7.21   Oil/water separator (for bilge arrangement)

In Figure 7.22 the oil/water sample is delivered at the high inlet, position 1, and then undergoes deaeration and an initial separation in the upper section of the machine, with the separated oil flowing directly into the oil recovery chamber at 14.

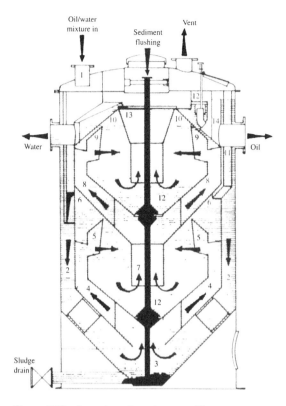

Figure 7.22 Operation of gravity type oil/water separator

The remaining partially separated mixture passes down through the annular space 2 to the lower part and into opening 3 to reach compartment 4. The mixture is then diffused in position 5 where much of the oil in the form of droplets is retained by a ring of vertical blades.

As these droplets coalesce and rise upward on the blades, their buoyancy increases sufficiently to cause the oil to float off the blade tips. This oil is then collected at position 6 to flow upward through the oil recovery column 11 towards the oil recovery chamber 14. The mixture left, which is now nearly oil free, is allowed to pass through the conduit 7 to enter compartment 8 where it is diffused again with the set of blades at position 9. A similar action causes residual oil to be collected at position 10 which is caused to flow upward to an adjustable weir at position 12 and then finally passed to the oil recovery chamber 14.

Any clean water left rises through the funnel 13 to overflow at the water outlet pipe. The level of the surface of the water above the lip of the funnel establishes a reference level above which the oil in the recovery column rises. This height is directly related to the difference in density of the oil and water. The oil in the recovery columns flows over weirs positioned higher than the water reference level and into the recovery oil

chamber 14. The weirs are set lower than the maximum height to which the columns will rise hence if oil or grossly contaminated water enters the separator then such samples would not pass down to the lower part but flow directly to the oil recovery chamber 14.

Incinerator operation

Where a ship is equipped with its own incinerator (see Figure 7.23) the responsibility for its operational safety is assumed by the Classification Society.

The basic requirements for such equipment are:

(a) The oil burning arrangements must be in accordance with the Society rules.
(b) The incinerator must not be used for waste having a flash point below 60°C.
(c) The loading door can only be opened when the firing supply is shut down.
(d) The induced draught fan is to be in use at all times when the loading door is open.
(e) The flue of the incinerator is to be independent of boilers and must be led to a suitable area clear of the accommodation and service spaces.

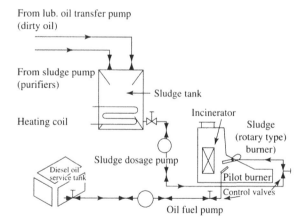

Figure 7.23 Diagram to show the operation of a typical incinerator

Permit to work systems

A Master is expected to have on board a 'permit to work' system in operation and that such a system should cover shipboard operations such as:

(a) Any electrical work.
(b) Hot work permit (for flame cutting, burning, welding, etc.).

(c) Overhead/overside working (e.g. mast or crane working).
(d) Entry into enclosed spaces (e.g. tank entry).

The purpose of having the system in place is to ensure the personal safety of those on board engaged in shipboard operations. It is also a system which ensures that the overall safety of the vessel is not put at risk.

The permit to work system ensures that a safe working practice operates on particularly hazardous operations for which, advance planning involves the co-operation of all parties concerned. It will generate communication either directly or indirectly with all persons involved in the proposed operation and will also monitor any guard systems which are in place for safe execution.

The control of an ongoing operation under a permit to work provides observation and close monitoring of the work being carried out. It ensures that the operational site is left in a clean and safe condition and that nothing will be left to endanger other persons or the safety of the vessel.

It should be realized from the outset that a permit to work will not in itself make any 'job' safe, but will provide a guide to operators and supervisors of the safe procedures required. It is issued in quadruplicate:

The original copy is held by the senior supervisor at a central control room.
The first copy is displayed at the working site of operation.
The second copy usually goes to the bridge/Master.
The third copy is placed in the ship's record file.

For additional reference see the Code of Safe Working Practice.

8 Offshore seamanship

Introduction

The last fifty years have seen many forms of offshore development worldwide. The recovery of raw materials from below the sea bed has increased subsurface and surface activity considerably. The obvious associated problems for the mariner in these areas have generated increased navigation hazards, greater workloads and the need for an ever growing requirement for prudent seamanship.

Traffic density alone for through vessels within this environment is of prime importance and with the increased number of obstructions, both surface and submerged, transit of such an area can be considered as 'high risk'. If the vessel is a working ship, like a tanker, actively engaged in the offshore operation, hazards such as fatigue and the deployment of resources need to be managed efficiently.

The command skills of the Master coupled with seamanship and navigation will be fully stretched on any vessel negotiating an offshore region. The restrictions on Masters to maintain their vessels within clear margins of safety are well documented and advice can be sought from many reference sources, i.e. *Mariners' Handbook*, navigational charts, routeing manuals, Annual Summary of Notices to Mariners, etc.

However, with the levels of technology, communications and legislation advancing at such a pace there is an ever increasing amount of knowledge needed by shipping personnel, not only to keep abreast of operational detail, but to stay within the letter of the law.

The offshore environment is changing with new ideas starting to encroach on what was previously just oil and gas recovery. The marine stage now caters for 'prison ships', while space exploration is being advanced with the 'Sea Launch' programme of satellites from mobile rocket launch platforms. New ideas are being generated constantly, and it would be inconceivable for mariners not to be involved when seven-tenths of the Earth's surface is water.

Passage plans through offshore regions

Many vessels will find their routes passing through offshore development areas and the Master is expected to advise on any specifics that need to be incorporated into the passage plan. Instructions to the navigator during the 'appraisal' stage of planning the passage should include attention to the following offshore references:

1. Notice 'Number 20' of the annual summary, with respect to safety zones around installations. (Recommended minimum passing distance of 500 metres unless otherwise informed.)
2. Cumulative rig position listing, Annual Summary of Notices to Mariners.
3. The *Mariners' Handbook*, section on offshore installations (NP100).
4. Large-scale charts, and the Chart Catalogue.
5. Routeing charts and recommendations from 'Ocean Passages of the World'.
6. IMO routeing manual for recommendations on respective tracks.
7. Weekly Notices to Mariners and in particular Temporary and Preliminary Notices.
8. Sailing directions for the area.
9. Admiralty List of Lights.
10. Tidal stream atlas and Admiralty Tide Tables.
11. Navigational warnings affecting the area.
12. Relevant MGNs, MINs and MSNs.
13. Local by-laws from respective territorial authorities.

In addition to the above further advice could also be obtained from Port or Pilotage Authorities regarding the movement of buoys, new dangers, developments or specific navigational hazards. Mobile installations or outward bound traffic could also be a useful source of current and updated information which could be relevant to the vessel's passage.

Watchkeeping offshore

The need for effective watchkeeping is essential in all marine environments, but especially in offshore regions which carry with them additional responsibilities and the need for positive awareness of such items as:

(a) Close monitoring of the vessel's position and respective underkeel clearance.
(b) Continuous lookout for multiple navigation hazards.
(c) Observance of recommended 'safety zones' around installations (minimum standards require 500 metres but guard vessels very often exercise a 2 mile restricted zone).

Figure 8.1 Standby vessel operating in close proximity to an offshore installation

(d) Extensive traffic movement involving specialized vessels, many operating under restricted ability to manoeuvre regulations and exhibiting relevant day/night signals.
(e) Passage plan contingency details, especially with reference to the use of anchors.
(f) The need for bridge team activity, with the use of manual steering, engine room operations, doubling watchkeepers, radar observation and specific manoeuvres.
(g) Specific recommendations regarding the routeing of the ship and the observance of fairways and traffic separation schemes.
(h) Those areas of specific interest where the Master is expected to take the 'con' of the vessel, e.g. focal points at the junctions of fairways.

(i) Regular monitoring of weather reports with particular attention to restricted visibility.
(j) Continuous monitoring of relevant communications for the area.

Offshore installations

Figure 8.2 shows the undersea pipelines of the link-up between the Elgin/Franklin interfield with the Shell Shearwater field in the North Sea, and illustrates the real dangers from the misuse of ship's anchors in offshore areas.

The field is situated 150 miles east of Aberdeen, Scotland, in depths around 50 fathoms, and additional navigational hazards must be anticipated from extensive surface traffic, including helicopter activity and vessels restricted in their ability to manoeuvre.

Offshore fixed installation: special features (production platform)

The illustration in Figure 8.3 is a typical offshore drilling and production platform, showing all the salient features.

Semi-submersible surface platform

The platform in Figure 8.4 is a semi-submersible surface type with a single anchor leg system (SALS) and a single well oil production system (SWOPS).

The extensive perimeter of activity around an installation is not limited to the surface, although surface traffic could well be wide ranging. The undersea connections of manifolds, risers, anchor patterns and pipeline connections could restrict maritime operations by other vessels. Close chart inspection is recommended with extreme caution on the use of a ship's anchors. Communications should be clear and immediate if Masters find themselves navigating in such an area with the need to exercise contingency action.

Offshore traffic: semi-submersible offshore installation

Figure 8.5 shows a semi-submersible offshore installation classified with Lloyd's Registry for operation with unmanned machinery space (UMS) and dynamic positioning (DP), the specifications of which are:

- Operational draught – 12.6 metres,
- Transit/survival draught – 6.4 metres,

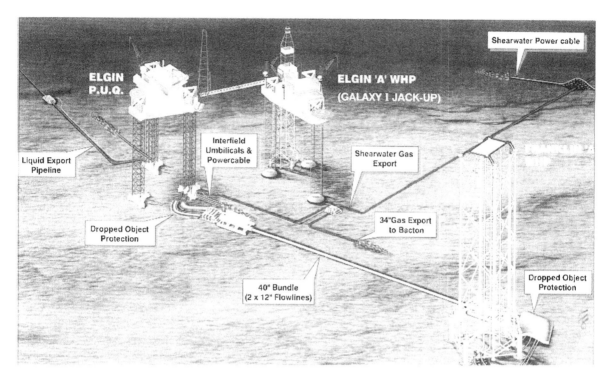

Figure 8.2 Undersea pipelines can prove a real hazard to ships when trying to anchor

- Main and auxiliary engines plus $4 \times 1250/1375$ kW azimuth thrusters plus $2 \times 1250/1375$ kW tunnel thrusters.
- Main cranage – 225 tons at 19 metre radius, 150 tons at 28 metre radius.
- Diving capability – 18 man system with two deck decompression chambers plus two × three man submersible decompression chambers.
- Complement – up to 89 persons, 3×38 man 'freefall lifeboats', heli-deck, and standard navigation/communication equipment inclusive of GPS, auto pilot, and Imarsat A.

The photograph in Figure 8.6 shows a 'jack-up' drilling rig under tow. The jacking legs are prominent with the drilling tower sited centrally and installation deck cranes, totally enclosed lifeboats and heli-deck can also be seen.

These structures are usually good radar targets but slow moving, and a wide berth is generally to be recommended. Such operations may be under the restricted ability to manoeuvre class. Figure 8.8 shows the installation in position.

Operations of offshore supply vessels (OSV)

The basic function of the supply vessel is to provide the necessary resources to maintain the installation in an operational mode. The operation of an OSV, however, is in itself not without hazard, because of the nature of its cargoes, the adverse sea conditions in which it is expected to operate, and the stability implications during and on completion of discharge. The low freeboards of these vessels and the possibility of icing taking place in certain areas could add further undesirable stability conditions.

In general the offshore supply vessel is small compared to the more usual commercial traffic and may or may not be fitted with its own cranes or derrick. Some vessels rely on the use of shoreside/installation-based lifting equipment. If the ship's own lifting equipment is employed, the vessel's centre of gravity is significantly raised when operating lifts in either the load or discharge operation. If this fact is coupled with the nature of other cargoes on board, i.e. fresh water, mud, cement or oil, there may be the need to take into account the effects of free surface in the stability calculations.

Typical OSV cargoes – drilling equipment, pipelines and the like – are known to retain large quantities of water (possibly up to 30 per cent of the volume of the cargo). With this factor in mind stability calculations should include an allowance, subject to the level of freeboard, of between 10 and 30 per cent. It is not unusual to see supply vessels with decks awash even in moderate weather conditions.

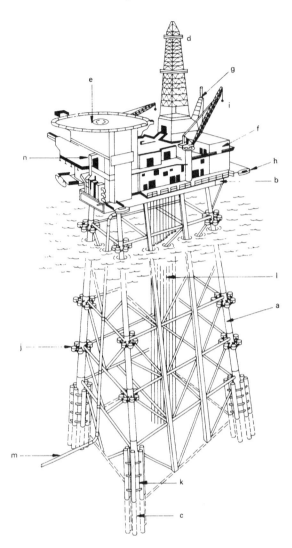

Figure 8.3 Typical offshore drilling and production platform: a – jacket; b – module support frame; c – piles; d – drilling derrick; e – helicopter pad; f – drilling and production equipment; g – flare stack; h – survival craft; i – revolving cranes; j – pile guides; k – pile sleeves; l – drilling and production risers; m – export pipeline; n – accommodation

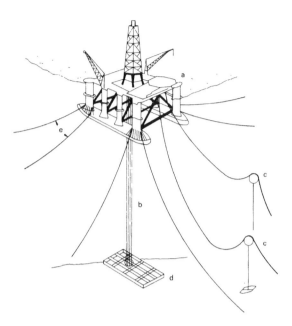

Figure 8.4 Semi-submersible surface platform showing: a–surface platform, b–multi-tube vertical drilling and production riser, c–flexible production risers, d–sea bed template, e–catenary moorings

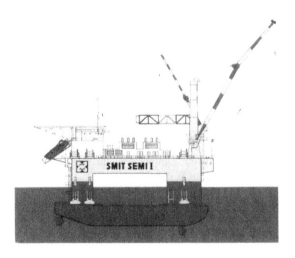

Figure 8.5 Semi-submersible offshore installation with UMS and DP

Stowage of deck cargoes aboard OSVs

Prior to loading deck cargo aboard the supply vessel, consideration should be given to the discharge which must take place from the top of the stow. Ballast arrangements should also be established before taking cargo on board, keeping in mind not only the loading and passage periods but also the discharge period.

Deck cargoes should not be stowed in such a manner as to restrict drainage towards the 'freeing ports' and/or scupper arrangements. Neither should cargo inhibit the closure of hatches, access points and other similar watertight enclosures. All cargo parcels should be securely lashed prior to departure to restrict any movement when the vessel is at sea. If open pipes are being shipped, the use of 'pipe plugs' should also be considered.

Figures 8.9 and 8.10 show typical offshore supply vessels and Figures 8.11 and 8.12 show the kinds of cargo OSVs are capable of carrying. Two types of cargo handling operation can be seen in Figures 8.13 and 8.14.

Figure 8.6 Mobile 'jack-up' drilling rig under tow in the Arabian Gulf

Figure 8.7 Shell 'Brent B' offshore platform. Drilling tower, heli-deck, flame tower, and deck crane are prominently identified

Requirements for standby vessels (SBV)

To meet the demands of the industry and the designated duties, standby vessels need to meet the following criteria:

(a) They should be highly manoeuvrable and capable of maintaining station holding.

Figure 8.8 Jack-up offshore exploration installation towed or self-propelled into position. Ongoing drilling operation is indicated by the pipe projection directly under the drill tower

(b) The bridge arrangement should provide an all-round visibility range of the waterline through 360°.

(c) The vessel should be equipped with at least two fast rescue craft (FRC), fitted with communications and standard recovery/rescue equipment. These craft should have the capability of being launched quickly while the SBV is underway.

(d) The SBV should have effective communications between the FRCs, the offshore installation, aircraft and the shoreside authorities.

(e) Different methods of recovery of persons from the sea should be readily available in addition to ladders or scrambling nets. (These require a degree of self-help by the would-be survivor and circumstances may be such that persons in the water could be injured and incapable of attaining a safe haven without an alternative system.)

(f) The SBV and the FRCs carried by the parent vessel should be equipped with searchlight capability.

(g) Coxswains and crews of FRCs should receive special training for duty with such craft.

Standby vessels are often engaged in a dual role and have been seen to be able to perform adequately alongside their primary duty of safety. However, whatever secondary role is engaged, it should not override the safety role.

Anchor handling

Offshore structures are positioned and secured by a variety of methods and an installation held by an extensive 'anchor pattern' is not unusual. The size of the structure, the depth of water and the environment influence the number of anchors employed but as many as ten

Figure 8.9   Multi functional O/S supply vessel showing: Heli-pad forward, fast rescue and totally enclosed boat stores crane aft, and triple bow thrust fitted

Figure 8.10   *British Emerald* engaged in offshore supply. Large open cargo deck space aft and low freeboard are regular features of supply vessels

Figure 8.11 Close manoeuvres by supply vessel to engage the installation's cranes prior to cargo discharge. Typical cargoes include pipeline, case goods, fresh water, cement, containers, aviation spirit, food supplies and general stores

Figure 8.13 Offshore supply vessel (OSV) engages in discharge to fixed offshore installation. Even the extensive outreach of the crane jib doesn't allow the OSV a great margin for manoeuvres

Figure 8.12 Containerized cargo – aviation fuel for helicopter activity, ready for dispatch to installation

Figure 8.14 Open deck of a supply vessel being employed for the dual role of anchor handling and supply. Currently employed in laying moorings and buoy positioning

anchors in a holding pattern are normal. These main anchors are very often backed up by either clump counterweights or piggy anchors (second supporting anchors on the same mooring).

To position a structure in this manner by the spread of outlying anchors would be preplanned following survey of the area. The order of running and positioning of anchors may also involve more than one anchor handling vessel and be conducted over a relatively extensive area.

Masters in transit through an offshore area should give such an ongoing operation a wide berth bearing in mind that the positions of the laid anchors may or may not be indicated by outlying position indicating/recovery buoys. Buoying arrangements will differ and could have intermediate 'spring buoys' beneath the surface in a position between the anchor and the surface buoy. This

system could further extend the range of obstruction from the installation and encroach on the course track of through traffic.

NB. Very few anchor marking buoys carry lights.

It is expected that vessels passing through such areas do so at an appropriate safe speed for the prevailing conditions and be engaged on manual steering. Lookouts should be briefed to report sightings of buoys as well as any specialized vessels. The use of own ship's anchors in such an area must be considered unwise and a contingency safe anchorage for the vessel should be identified and included in the passage plan at the onset.

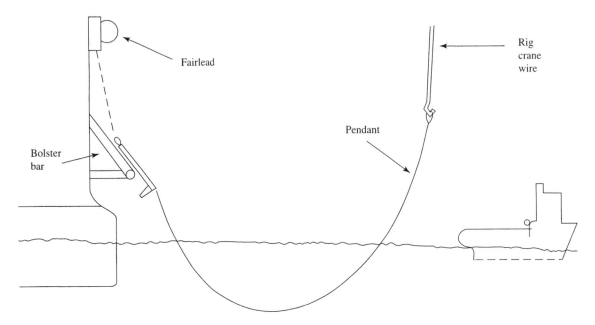

Figure 8.15   Laying anchors. The anchor is walked back to a position where the shank is resting on the 'bolster bar' and the flukes are clear. The 'anchor pendant' is then lowered to the anchor handling vessel to be secured to the towing spring

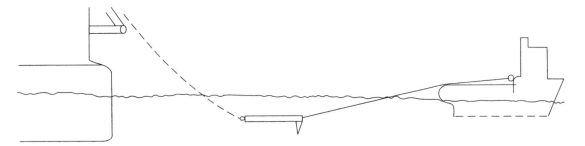

Figure 8.16   Walking back. The slack on the towline is taken up and the anchor is walked back. The weight of the anchor is then hauled out towards the stern of the vessel

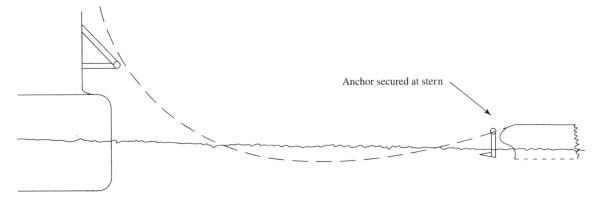

Figure 8.17   Hauling out. The anchor handling vessel continues to haul out the anchor and cable to the holding position, keeping a shallow catenary on the chain

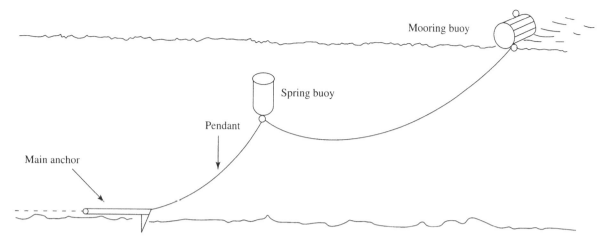

Figure 8.18 Lowering and positioning anchors. The anchor is lowered and the pendant is buoyed direct to a mooring buoy, or to an intermediate spring buoy and then to the mooring buoy

Anchor handling operations can take a considerable time and commercial vessels in the area should have consideration for their own well-being as well as for those persons engaged in laying or recovering anchors. Local communications could well provide the most up-to-date information on such an activity, in addition to navigation warnings. Close monitoring of weather reports, especially the state of visibility, must be a priority for a prudent Master if the ship's track is unavoidably placed in the proximity of such an operation. Further communications may be sought from the installation itself and/or from the anchor handling vessel regarding their perimeter of operation and location of specific obstructions.

Figure 8.20   Anchor buoy with recovery strops prior to release from the stern of the anchor handling vessel. The anchor pennant can be seen passing through the centre of the buoy and secured at the upper shackle

Use of 'piggy anchors'

Some operations use secondary 'back-up' anchors known as 'piggy anchors' to provide additional securing for the main anchor holding. Masters should note that this system can extend the length of underwater chain by up to a further 300 metres from the main anchor position, depending on the size of the project.

Offshore specialized traffic

The needs of the offshore industry continue to grow with increased movement into deeper water and developments

Figure 8.19   Anchor positioned prior to letting go on the stern of an anchor handling vessel. Recovery buoy pennant is seen extending forward above deck area

into areas that were previously thought to be non-commercial. With such expansion taking place on a worldwide scale, there has been a need to maintain maritime support by introducing many new styles of operational vessels. Such developments have meant that ship's Masters, officers and crews have been forced to diversify their activities and become familiar with the operation of diving support vessels (DSVs), remote operated vehicles (ROVs), dynamic positioning (DP), cable/pipe laying activity, helicopter operations and fast rescue craft (FRC), to name but a few.

The senior officer with DP qualifications or cable laying vessel experience is no longer a rarity. The skills of these mariners, especially those associated with the offshore industries, have increased considerably in a variety of fields over the last 30 years a fact which has broadened the horizons of all seafarers by creating an awareness of the intricacies of a sister industry. With this increased knowledge being introduced to the ship's bridge, as well as the engine room, it must be expected that increased harmonization between personnel will take place, and individuals will be conscious of each other's activities.

Dynamic positioning

Dynamic positioning is the term used to describe a vessel's ability to hold station by means of position referencing. It is a concept which is extensively employed on specialized vessels within the offshore industry aboard offshore supply vessels (OSVs), diving support vessels (DSVs), cable laying or other similar survey vessels.

In all examples the vessel so equipped holds station using automatically or manually controlled thrusters. These thruster units directly influence the six degrees of freedom exercised on the vessel (see Figure 8.21).

Station holding by dynamic positioning methods is generally achieved by one or more of the following three position referencing systems (PRS).

Taut wire (PRS)

This method employs a heavy weight of approximately half a tonne being lowered to the sea bed on a wire via an extended boom. The wire is directly linked to a constant tension winch usually set at about 0.25 tonne.

The length of the wire extended, together with the angle to the vertical can be monitored in both the longitudinal and transverse planes by sensors at the sheave and the winch. The position of the ship can then be defined from the data relayed back from the sensors into the DP system.

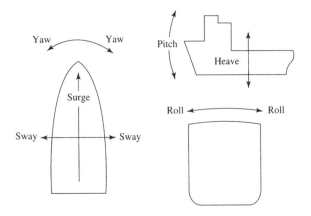

Figure 8.21   The six degrees of freedom

Hydroacoustic (PRS)

This method is achieved by a transducer situated on the bottom of the vessel and transponders positioned on the sea bed. A signal is transmitted to the transponders and then returned to the ship in a similar way to echo sounding. The range and direction to the transponder can then be ascertained defining the ship's position in relation to the transponder and relayed into the DP console.

Artemis microwave (PRS)

The Artemis system is achieved by a radio link between two transceivers, one positioned on the ship while the other is situated on a fixed reference point, like an offshore installation. The microwave link between the two can be calibrated to define range and bearing from the fixed reference which is then relayed to the DP system.

If drift from the position occurs by external forces, thrusters can be automatically engaged to correct the holding pattern. DP operators could also provide manual override in a variety of circumstances to suit specific operations. e.g. diving support work.

Watchkeeping duties with DP

It is considered normal practice for vessels operating with dynamic positioning (DP) to have two Watch Officers – one designated the DP watchkeeper, while the other carries out the more usual Officer of the Watch duties.

When taking over the watch the designated DP officer is expected to confirm the ship's position and make himself/herself aware of the ongoing work affecting the vessel's situation. The officer would also clarify the

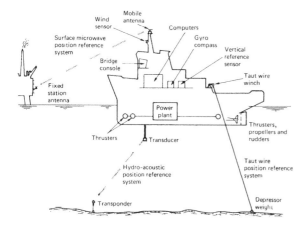

Figure 8.22 Installation features of dynamic positioning

position referencing systems in use or on standby and confirm all internal/external communications.

Overall performance is monitored by the DP officer, and the power supply, together with all alarm systems, would be assured. A chalk board is generally employed to provide continual updated status as to which elements are engaged at any one time, i.e. active generators, on-line thrusters, etc. A DP log book should also be maintained.

Example standing orders for DPOs

1. The DP officer must not be relieved while ongoing manoeuvres are in progress.
2. A minimum of two position referencing systems must be employed when engaged in station holding within 100 metres of any obstruction.
3. A minimum of three position referencing systems must be employed when the vessel is engaged in 'diving support' activity.
4. Operational checklists must be complied with in full, prior to commencing any specialized engagements.
5. The vessel must be observed to be on station and steady for a minimum period of 30 minutes before commencing any activity.
6. The alarm systems and capability graph must be checked and defined.
7. Contingency and escape plans must be established before commencing station holding.
8. The Master must be called at any time where the officer is in doubt regarding the operational status of any element of the DP system, or if the vessel is not holding her position as desired.

Figure 8.23 shows an example of capability 'footprint', used when station holding with DP.

Tanker operations

Use of offshore floating storage units (FSUs)

Many tanker personnel will be familiar with making a pipeline connection with the many types of offshore storage units currently operational around the world today. Variants in mooring arrangements abound with many vessels making either a semi-rigid connection, where an 'oil boom' supporting the oil bearing pipe will be deployed either from the stern or the bow of the storage unit to the carrier vessel, or alternatively, soft moorings with chafing (securing) chains and coupling to a floating, flexible oil pipe (see Figures 8.24 to 8.31 for a variety of working options and equipment used).

Mooring operations

The mooring of any vessel is a hazardous operation, especially in the oil industry where the sheer weight of equipment items has to be considered. Light messengers if 'snagged' could easily part with the more heavier hawsers and/or chain arrangements causing serious damage or personal injury. Mooring rope thimbles, shackles and chafing chains are of considerable size and very awkward to handle, and Masters need to make crews aware of the necessity for even basic precautions – hard hat use, warm and heavy duty protective clothing, reinforced footwear, and the dangers of wearing jewellery or loose clothing.

Offshore navigational dangers

Manoeuvring large vessels like VLCCs in the proximity of offshore installations, floating storage units or any of the many types of well heads is a navigational task with its own particular hazards. Clearly the need for a slow manoeuvre, which may take into account considerable small boat activity and the associated dangers of 'interaction' with such craft in close proximity, is essential and is especially noticeable when line of sight from the ship's conning position is lost as small craft move under the flare of the bow or become obscured around the stern.

Masters in general will be obliged to 'con' the vessel from the bridge position which in many cases is a long way aft from the mooring operation at the forward end. He will often be totally reliant on clear and effective communications from the officer in charge of the mooring operation. Such situations call for direct ship control to the forward end and many vessels have

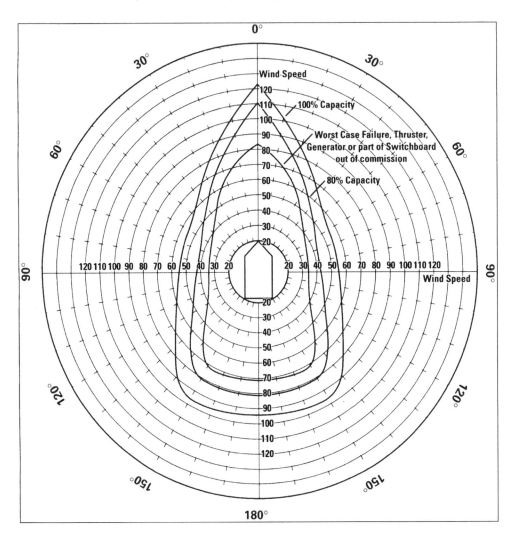

Figure 8.23 Example of capability 'footprint' for a vessel when station holding with dynamic positioning

been adapted to provide hands-on control of the vessel from a forward conning, vantage position. Such an operation requires the transfer of controls of engine, propellers, helm, and thrusters to effect station holding while moorings are established and pipelines landed and coupled.

The number and type of traffic in and around an offshore development area can be considerable. Many vessels may be involved in specialized activities such as diving operations or helicopter reception which could further restrict their ability to manoeuvre and cause the display of restricted navigation signals.

The bridge team would need to be conscious of the fact that these activities place the responsibility of giving way onto the shoulders of vessels which are not restricted or hampered and keeping sufficient sea room available may prove a prudent action.

Figure 8.24 Large tanker moored bow to stern of floating storage unit. Oil boom being deployed from the FSU supports oil bearing pipe to be connected to the tanker's forward manifold

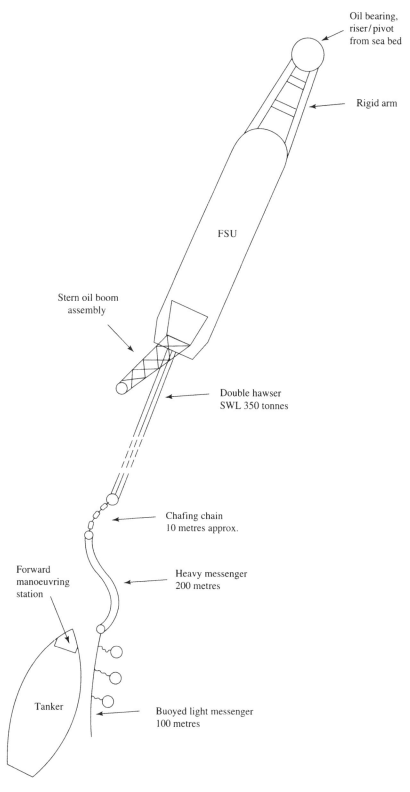

Figure 8.25   Mooring example to floating storage unit (FSU)

Figure 8.26   Tanker moored to SBM – topped derrick for hoisting flexible (oil bearing) flotable pipeline

Figure 8.27   SBM with single mooring arrangement and flexible oil pipe illustrated

Figures 8.32 to 8.36 show different mooring arrangements and equipment used.

Cable laying activity

Introduction

One of the largest proactive maritime activities today is without doubt the submarine cable industry. Communications worldwide have expanded and the use of fibre optic, has moved alongside the internet to make the information highway accessible to all. Alongside such growth has been and continues to be the marine cable laying activity.

The vast amount of equipment and resources needed to conduct cable operations is impossible to cover in this text and therefore a general overview is presented here.

Several large companies are heavily involved in submarine cables and with cable networks linking many countries a great number of personnel are directly involved in the industry both afloat and in support ashore.

Once cables are laid they must be maintained and in order to lay, recover and inspect cable lays, specialized vessels are employed within the marine environment to carry out the specific tasks generated from such a complex industry. Navigational accuracy for such operations is an absolute necessity and to this end many vessels within this industry employ dynamic positioning to ensure precision station holding for cable activities.

Cable links

There are many cable networks already established worldwide, while many more are still under construction. To provide a general background the following examples in the Caribbean Sea, and from UK to Japan are used.

Eastern Caribbean fibre system

This system is the longest unrepeated inter-island link and runs between Trinidad and Tortola in the Virgin Islands. To establish such a link requires detailed planning, effective surveying and careful route selection. Designated landing sites where cables come ashore need to be suitably vetted and agreed between local authorities as well as being physically and geographically appropriate.

Total length of cable link 1729 km. Longest span 285 km.

Cable operations, although included within 'offshore activity', are of course worldwide with links established across large areas of the oceans including the Atlantic. However, by the very nature of the industry much of the work is conducted close in shore where marine cables tend to meet at a 'beach manhole' and join with the land cable, i.e. 'beach joint'.

Such operations generate considerable activity and high traffic density, similar to any other offshore region. The ships themselves are often restricted in their ability to manoeuvre, for reasons other than the cable laying activity. Many cable ships operate remote operated vehicles (ROVs) for cable inspection and the majority have purpose-built heli-decks and could well be involved in launching or recovering aircraft.

Vessels involved in cable laying were mainly constructed to conduct their activity from the bow, but more recently stern-working cable layers have become operational, e.g. *Cable Innovator* a purpose-built (1996) stern-working cable layer, part of the Cable & Wireless Global Marine fleet.

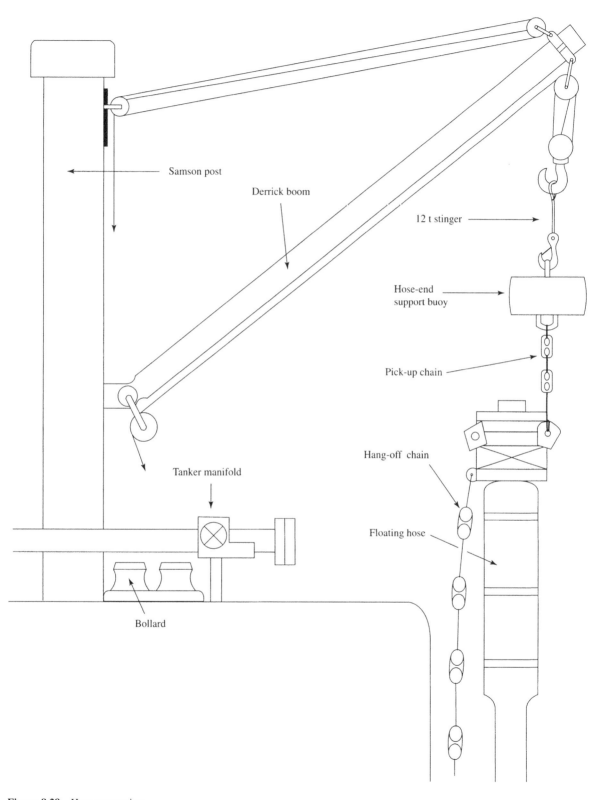

Samson post

Derrick boom

12 t stinger

Hose-end
support buoy

Pick-up chain

Hang-off chain

Tanker manifold

Floating hose

Bollard

Figure 8.28 Hose connection

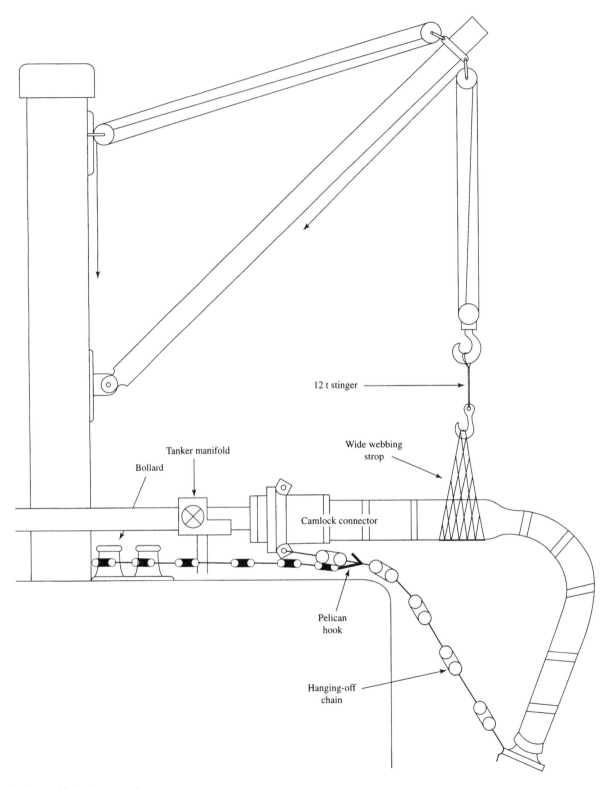

Figure 8.29   Hose connection

Figure 8.30 Chafing chain and lead to SBM

Figure 8.31 An auto kickdown (AKD) bracket employed to secure chafing chain

Figure 8.32 Chafing chain – recovery buoys being employed prior to securing to onboard stopper arrangement. Rigging/mooring launch assisting alongside floating oil transfer pipe

An example of a typical cable laying vessel in dry dock can be seen in Figure 8.39 and Figure 8.40 shows the vessels that make up the Cable & Wireless cable laying fleet.

Porthcurno to Miura cable system

This cable example is operational from Porthcurno (UK) to Miura (Japan) over a distance of 26 770 km. It passes via Suez and has at least 15 landing points in more than 10 countries.

In establishing the lay of cable several different burial tools were used from specialized cable laying barges, one of which is shown in Figure 8.42.

Cascade™ (computer assisted slack control and cable deployment)

The *Cable Innovator* was involved in 81 days' continuous cable laying activity and used the 'CASCADE™ system' for the first time on the Gemini North and Gemini South cables between the USA and the UK at depths of up to 1330 m.

The Gemini Project was a joint venture between WorldCom and Cable & Wireless.

The CASCADE™ system provides real-time modelling of the cable distribution and reduces cable slack on the sea bed. This effectively cuts wastage of cable and reduces the risk of 'chafing' or 'hooking' during cable laying operations (see Figure 8.43).

Glossary of terms used in the submarine cable industry

Articulated piping Fitted over cable lines in shallow water to provide more protection than that afforded by amouring alone.

Beach anchor Strong point embedded in the ground which acts as a holding anchor for the shore end of the cable.

Figure 8.33 Mooring arrangement between the FSU and a VLCC – chafing chains being lead from panama leads set into the bows of the VLCC. Floating oil bearing pipeline visible between the two units

Figure 8.35 Forward mooring arrangement showing panama leads and bow stoppers for accommodating the chafing chains

Figure 8.34 Bow stopper to accommodate chafing chains

Figure 8.36 Oil bearing pipe coupling between the VLCC and the FSU

Figure 8.37 *Seaway Falcon*, built in Holland as a drill ship and converted into a cable laying vessel, and fitted with Simrad dynamic positioning system. Capability: two stowage holds, for flexible cable below decks, one carousel, and six reels of flexible pipe on deck. Total cable capacity 9000 tonnes

Beach point The position of joining where the land cable meets the shore-end cable at what is known as the 'beach manhole'.

Beach manhole A void space positioned above the shoreline where the sea cable and the land cable are joined.

Bell mouth A circular opening above the cable tank which allows the cable to uncoil freely from the tank towards the deck.

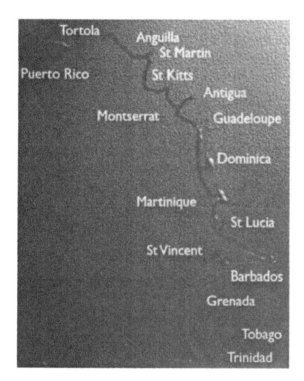

Figure 8.38 The Caribbean fibre system, linking Trinidad and Tortola

Figure 8.39 Cable laying vessel, seen from the aft end, when in dry dock. Cable sheave and lead arrangement for stern operations are visible. Bumpkin framework protects the rudder and propeller region

Bollard pull When related to cable ships refers to the vessel's ability to tow a cable plough.

Bow baulks The forward supports of a cable ship which contains the cable sheave over which the cable is recovered.

Bow to stern transfer An operation which describes the movement of the cable from the bow sheaves to the cable machinery at the after part of the vessel.

Employed by a conventional 'bow-working' ship at the start of the lay when a previously laid cable end is picked up at the bow position and passed around the outside of the hull to the after end.

Bumpkins Guard frames positioned either side of the vessel's stern to keep cables and ropes clear of the ship's propellers during cable/transfer operations.

Cable tank A circular storage area in the ship used for cable stowage.

Cable drum A powered drum employed for recovering and laying out cable.

Cone A truncated conical structure situated in the centre of the cable tank which prevents the cable being stowed at less than its minimum bending radius and also holds the stowed cable in position.

Conventional coiling A method of coiling cable clockwise into the cable tank working inwards from the perimeter of the tank towards the cone.

Crinoline A framework around the cone which prevents any whip motion on the cable if the height from the top of the stow to the 'bell mouth' is great. It can be adjusted in height to suit the lead on the cable.

Cut and hold grapnel A mechanical grapnel capable of cutting the cable and holding the end to facilitate recovery back to the ship.

Cutting flatfish A grapnel with a diamond shaped plate with a prong set each side and a blade fitted at the junction. Once the cable is lodged the cutting of the cable can be carried out.

DOHB Draw Off, Hold Back. A cable engine which is used in conjunction with the cable drum to prevent the cable slipping around the drum.

Double armour A type of cable protection consisting of two layers of armour wire. Used extensively at the shore end of the cable.

Egglink An egg-shaped steel link for cable working and chain fitting.

Equalizer A passive device used to match cable loss and repeater gains throughout an oceanblock within an analogue cable system.

Final splice The final joint in a cable repair operation which returns the cable to operational status.

Gifford A grapnel of four wide-seated hooks set at right angles to each other for use on hard or rough sea beds.

Grap rope A 6×3 wire core rope used for grappling.

Grap sheet A large-scale chart employed to plot a ship's position whilst engaged in grappling for a cable.

Holding drive A grappling drive used to recover cable as opposed to cutting.

Klick Term of measurement when laying cable, equivalent to one kilometre.

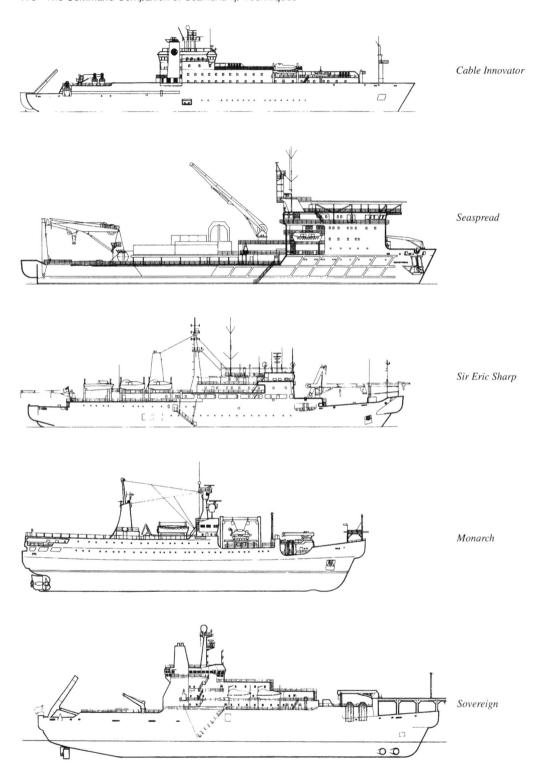

Cable Innovator

Seaspread

Sir Eric Sharp

Monarch

Sovereign

Figure 8.40   Profile of cable laying vessels of the Cable & Wireless fleet

Figure 8.41 Route taken by the Porthcurno to Miura cable system

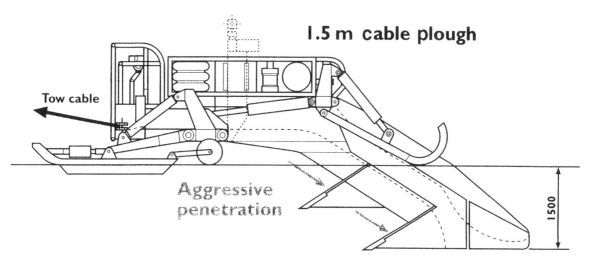

Figure 8.42 Cable ploughs are currently designed to operate up to a depth of about 2000 m and trench to a depth of approximately 3 metres

Layback That distance from the stern of the vessel, (or the principal point of navigation) to the position of a towed vehicle or that point at which the cable touches the sea bed.

Linear cable engine A series of wheels situated at the aft end of the vessel. The wheels are fitted with tyres to grip the top and bottom of the cable to bear the weight when picking up or paying out the cable.

Maul The term given to the damaged section of a cable when it has been struck by a fishing boat's trawl gear or a ship's anchor.

Mushroom Anchor An anchor formed in a mushroom/umbrella shape generally employed as a buoy mooring anchor (see *Seamanship Techniques*, Part 1, Chapter 2).

Normans These are pad eyes welded onto the upper part of cable buoys to accept the girth wires, or the buoy trailers, depending on the type of buoy rig.

NOTS Nominal Operating Tensile Strength, the tension that can be applied over a 48-hour period to optical fibre cable when in suspension with a greater than 95 per cent fibre survival probability.

NPTS Nominal Permanent Tensile Strength, the tension that can be left in optical fibre cable for its entire life span with guaranteed fibre survival.

NTTS Nominal Transient Tensile Strength, a term that is applicable to fibre optical cable which gives the tension that can be applied for a one-hour period with a 95 per cent probability of survival. (Usually expressed in kilonewtons.)

Figure 8.43   Detailed information about the sea bed geography allows the optimum deployment of the cable and minimizes the risk of unwanted suspensions which could affect the long-term integrity of the cable

OOS Out of Service, the term used to describe a cable which is still in situation but no longer in service.

Pipe tracker A device used for locating buried submarine cable and usually works in conjunction with an ROV.

Power Safety Officer That designated person aboard the vessel who is responsible to the Master for the safety of personnel with respect to dangerous voltage and current when testing and repairing systems and operations.

Pre-lay grapnel run Route clearance of cables and debris prior to a cable lay.

Preformed stoppers Standard size wire helixes which are designed to wrap around and grip cables allowing a transfer of weight from the cable's inner wire rope to an external grip, without causing damage to the outer sheathing of the cable.

Protection grounding unit This provides an earth path in the event that a cable becomes live during a repair which has switchable branching units. It allows

ongoing repairs on a leg of a system without loss of traffic on other branches.

Rigging the buoy Preparation of a buoy with rope or chain, prior to use.

ROV Remote Operated Vehicle. A submersible tool usually powered and controlled by the mother ship on an unbilical cord. Used extensively for subsea cable inspection.

Reverse gifford A gifford grapnel used backwards in conjunction with other grapnels to act as ballast.

Sheaves Large grooved wheels situated at the bow and/or stern of the vessel to accommodate recovery or laying of cable.

Splice The joining together of wires or cables.

Stern to bow transfer The movement of the cable from the stern machinery to the bow sheave, as performed by a conventional bow-working vessel when making the final splice.

Weather window A time between two forecast periods of bad weather, when operations can be conducted.

9 SOLAS: revisions, amendments and developments

Introduction

Those who possess a Master's qualification should be all too familiar with the standards of fire-fighting and life saving appliances on board the vessels of the Mercantile Marine. It is not the function of this chapter to reiterate basic seamanship but to update the mariner with changes and innovations which have taken place alongside increased levels of technology.

New tonnage has created new problems for the Master; high sided vessels present different challenges when recovering persons from the water, for instance. The new design of the catamaran hull (Figure 9.1), and the use of alternative survival craft/systems as opposed to conventional lifeboats, all provide the modern Master with food for thought.

Evacuation and abandonment

Many Masters have never had to put emergency procedures into operation. However, history has provided many examples of just why we need to be well practised for the unexpected eventuality. The dangers of complacency, lack of training, incorrect decisions, faulty equipment, incompetence, or just plain bad luck can be attributed to many seagoing disasters. Hopefully, many lessons have been learnt from such incidents as the *Estonia* (1994), *Herald of Free Enterprise*, off Zeebrugge (1987), and the *Sally Albatros* (1994), where a marine evacuation system (MES) was employed to evacuate 1250 passengers successfully from a Finnish cruise ship, aground in the Baltic.

In the above examples large numbers of passengers were involved and even with all the modern, up-to-date facilities, the industry has suffered major loss of life. With current building, the cruise industry is enjoying a boom period with larger vessels being constructed to carry in excess of 3000 passengers. The associated problems of evacuating large numbers of persons from

a ferry or a Class 1 cruise ship are considerable for the Master. He or she must take the 'con' of the vessel in a position on the bridge in order to control and monitor communications and the essential navigation of the vessel. This situation forces duties to be delegated possibly to the less experienced officer who may not have been trained in crowd control.

The time when essential training comes to the fore is during an emergency, and in particular if the situation develops into an abandonment of the vessel.

Figures 9.2 to 9.5 show the types of liferaft fitted to vessels and methods of launching and evacuation.

Masters cannot take on all the shipboard duties during an emergency; they must delegate and trust that persons in charge of survival craft can exercise management skills as well as seamanship skills, when acting as a coxwain of a survival craft. With increased tonnage sizes and increased passenger capacity (Figure 9.6) life saving appliances have had to cater for larger numbers of persons and marine personnel will have to learn to deal with these larger numbers if caught up in an emergency situation.

Crowd control for larger (passenger) vessels

The type of crowd control required on board a ship is one with the common motive of mutual survival. This is not to say that panic will not be generated in an onboard emergency. Being well prepared to manipulate large numbers of passengers is half the battle if an evacuation from the vessel became necessary. Master's are always conscious of the fact that the mother ship provides all life support systems and would not be abandoned unless it was no longer sustainable. However, once the situation deteriorates any decision to evacuate passengers and crew should be made sooner rather than later, in order to provide an orderly manner of disembarkation.

Cruise ships and high sided ferries may have to lower survival craft down to a waterline which may be in

Figure 9.1   New hull design. Catamaran twin hull ferry operates with 432 passengers and 85 cars. Currently engaged with the Isle of Man Steam Packet Company on the Irish Sea short ferry trades

(a)

(b)

Figure 9.3   Davit launched liferaft operation

Figure 9.2   Abandonment training exercise into liferafts

excess of 20–30 metres, possibly at night and possibly in bad weather. This in itself is no mean task for even the most experienced of seafarers. If passengers cannot be organized in advance, managing the situation during an ongoing incident may become impossible.

Alternatively, deploying a marine evacuation system (MES)*, see Figures 9.7 and 9.8, is only the first stage of such an operation. The movement of many persons down the slide could generate crowding at either the embarkation point or on the lower raft embarking platform.

NB. The regulations require that one man is capable of deploying the MES, but Masters should be aware that those personnel required to launch the liferafts may also

---

* In the case of the *Herald of Free Enterprise*, which capsized onto its side in shallow water, the MES could not be deployed, neither could boats from davits be safely launched. Throwover liferafts, lifebelts and lifejackets backed by helicopter support became the order of the day.

be needed to relieve the potential survivors from the integral embarkation platform. Also the vessel needs to be in an upright or nearly upright condition for the MES to become operational.

Marine evacuation (chute) systems (MEC)

Chute marine evacuation systems are probably the latest innovation to enter the marine environment. Different manufactures have incorporated a variety of ideas in establishing chutes as a viable means of abandonment, especially important to high sided vessels and currently being introduced to exceed the self-righting requirements for liferafts of roll on-roll off vessels.

The concept of the RFD system is to transfer evacuees directly into enclosed liferafts without exposure to the weather. Whereas the 'Selantic escape system' permits the transfer of would-be survivors to an inflated platform (like a standard MES) and thence to peripheral liferafts.

The Marin–Ark RFD system employs 100 person, self-righting, totally enclosed liferafts. Chute stations are

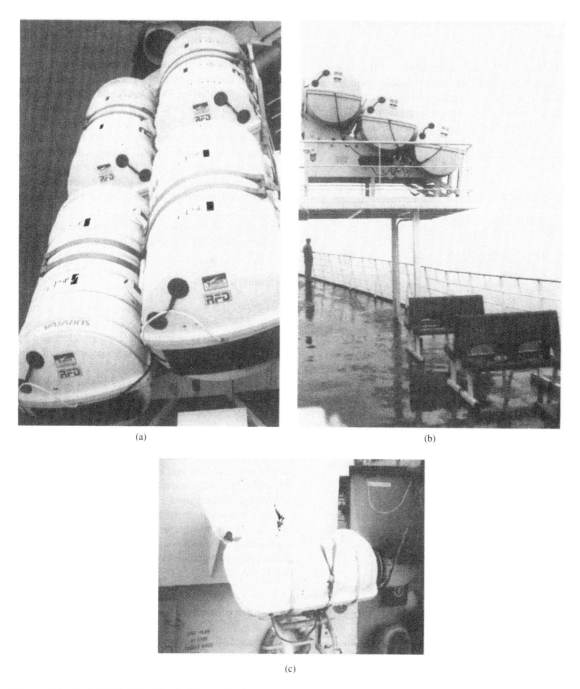

(a)

(b)

(c)

Figure 9.4 Racked inflatable liferafts for easy deployment overside would necessitate persons disembarking down the ship's side ladders and/or entering the water

positioned either side of the vessel which incorporate telescopic, double, near vertical chutes which act as the conduits to the survival craft – a single point activation which deploys the rafts, causes automatic inflation at the surface, on the end of the chute, once the release mechanism for the system is made operational.

Capacity of the system can be customized to suit the needs of the vessel. Operating between 8 and 17 metres in height with two chutes and four liferafts provides a 430 person survivor capability. An alternative example operates with two chutes and two liferafts providing a 212 person capacity (see Figure 9.9).

Marine safety evacuation chute (MEC): system operation

The DBC Marine of Canada manufacture a MEC system (see Figures 9.10, 9.11 and 9.12) which employs a fabric chute stored in a compact housing situated either internally or externally at the embarkation level. The system is deployed by a single handle release which automatically allows the chute to fall by gravity to the water and inflate the boarding platform.

Figure 9.5   A Viking marine Evacuation slide deployed with the lower disembarkation platform starting to fill with occupants. Forty-five persons liferafts are employed for survivors and the person in charge of a survival craft is expected to control the well-being of a fully loaded liferaft in possible adverse conditions

Figure 9.6   The *Oriana* Class 1 passenger cruise liner berthed port side to. She is the flagship of P&O Cruises operating with a crew of 760 and carrying just under 2000 passengers. There are extensive side cabin accommodation and a forward viewing facility; note the position of survival craft and embarkation deck on starboard side

The system works in conjunction with a range of reversible and self-righting liferafts from 25 person to 150 person capacity. Once the chute is deployed two platform crew would descend to secure the liferafts to the platform and monitor the throughput of passengers from the chute exit towards the liferafts. Once the liferafts are loaded the rescue boat tows them clear of the platform so additional liferafts can be deployed and loaded.

Crew control the evacuation procedures from the vessel and as the system is passive once inside the chute no effort is required to drop to the surface platform where the chute has a braking system to ensure the body is not exposed to a high stopping force. Evacuees are protected throughout the abandonment by the inner and outer material of the chute which is manufactured in a heat, flame and smoke resistant material.

Marshalling points

Under the SOLAS regulations passengers have instructions in their cabins to indicate where their emergency muster station is. These stations are clearly identified by poster displays together with directional arrows of how to reach the station.

On board large passenger vessels, muster stations are generally situated in the public rooms such as lounges, cinemas, etc., which can accommodate large numbers of passengers. They are usually positioned within easy reach of the embarkation deck for the survival craft. But how are passengers moved from these muster areas to the embarkation deck and into survival craft? Launching officers and the Master should note that 1500 persons on the average embarkation deck of a typical cruise liner will leave very little deck space available for operations like davit launched liferafts. With passenger numbers approaching 3000 in a confined area, working space and crowd control could become essential compatible factors (see Figure 9.13).

Passenger movement in an emergency

It is suggested that passengers be moved from the muster points in manageable numbers. On the assumption that manpower aboard ship is limited, a single person should be able to control a group size of approximately 25–30 persons and deliver them to the person in charge (coxswain) of a survival craft. Another factor in determining the number of persons to be moved is the size of the survival craft. Liferafts of the davit launched type accommodate 25 persons, while lifeboats are often of a 50 or 60 man capacity (suitable for multiple groups of 25–30).

Groups of this size could also be monitored as to who is in the boat/liferaft and who has yet to go.

Figure 9.7 Inflated evacuation slide, manufactured by Liferaft Systems Australia Pty Ltd. Evacuation times are expected to reflect 100 persons inside 5 minutes with two crew members assisting the operation

Figure 9.8 100 person liferaft supplied as a separate survival craft or in conjunction with the MES in Figure 9.7. Manufactured to meet the SOLAS 1986 requirements and tested in accordance with IMO resolution A689 (17) November 1991. The example shown has four access points, one at each end and an enlarged entrance on either side

Groups in excess of 30 would without doubt generate a blockage into enclosed boats or partially enclosed boats (see Figure 9.14) because of the limited size affecting access into survival craft, especially so if injured parties are being accommodated as well.

Special requirements for personnel on ro-ro passenger vessels

For additional reference see the revised STCW convention and the adopted resolutions, in particular Chapter V.

Training in crisis management and human behaviour is being examined by IMO, with the view to developing additional requirements for any person having responsibility for passengers.

Personnel designated on muster lists to assist passengers in an emergency situation will be required to complete training in crowd management and safety procedures, including:

(a) an awareness of life saving and control plans,
(b) the ability to assist and control passengers to muster points and then to embarkation stations,
(c) mustering procedures.

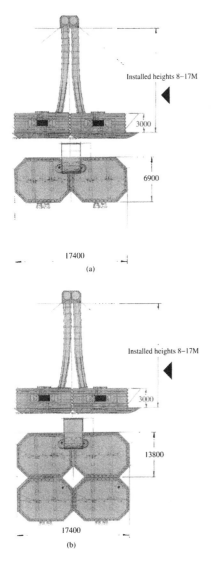

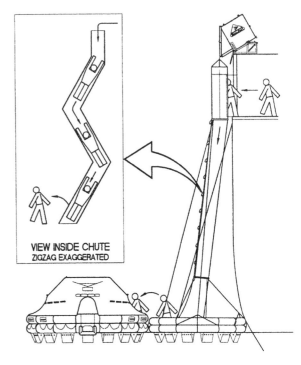

Figure 9.10 MEC with inflated surface platform and liferaft. Duration of fall 2–3 seconds from time of entry, for a 13 metre descent height

Figure 9.9 (a) Two chutes + two liferafts, 212 person capacity. (b) Two chutes + four liferafts, 430 person capacity

Extra familiarization training

Officers and other designated personnel who are assigned specific duties must receive specific familiarization training relevant to the ship, e.g. the operation, design limitations and procedures for the securing of 'hull openings'.

Figure 9.11 MEC II double chute deployed with British Columbia Ferries

Safety training for personnel

This provides a direct service to passengers, and personnel will have to be able to communicate with passengers with more extensive criteria than similar personnel on other types of ships.

Passenger safety, cargo safety and hull integrity training

Any personnel with immediate responsibility for embarking passengers or loading cargo will be required to undergo training appropriate to their duties to ensure that they can apply special procedures relevant to ro-ro passenger vessels.

Figure 9.12 Single chute deployed showing side fender and surface platform

Crisis management and human behaviour

This is currently under development, but the revised convention establishes the need to involve a degree of psychology training for persons responsible for the safety of passengers. It would also probably include additional skills in communications, stress handling, emergency planning and leadership qualities.

Incident report: Norwegian ferry *Prinsesse Ragnhild*

On 8 July 1999, while on route from Kiel towards Oslo the passenger ferry *Prinsesse Ragnhild* experienced an engine room fire. The Master transmitted a distress and evacuated over 1100 passengers. A marine evacuation system (MES) was deployed and response to the evacuation by local shipping was extensive, including fishing boats, other ferries and cruise ships in the area.

Additionally, nine helicopters assisted from Norway, Sweden and Denmark conducting winch operations to get people to safety.

A hundred and seventy-two crew members were later praised for their calm and efficient handling of the emergency, transferring the majority of passengers by lifeboats to other vessels. The fire was subsequently brought under control in 3 hours.

A seventy-three year-old lady later died from a heart attack and another 13 persons were treated for smoke inhalation.

Figure 9.13 Landing/disembarkation platform being tested to accommodate 80 persons. All persons wearing immersion suits and lifejackets illustrate how problems could be caused with controlling large numbers in confined areas

This incident was notable because of the massive response that attended the distress, nearly 20 vessels in all. The weather was also good. The length of the passage was only 19 hours yet a major engine room fire occurred highlighting that in the marine industry emergencies strike at anytime.

Lifeboat release systems: 'on load, off load'

Over the last decade several accidents, some fatal, have occurred with the release gear fitted to survival craft. It became clear, following investigation of these incidents, that all too often personnel responsible for the operation and maintenance of release mechanisms were not fully conversant with all the working aspects of particular release gear systems on board. A lack of understanding of the difference between 'on load' and 'off load' came to light.

Off load release

This is defined as a release system which requires all the weight to be taken off the lifeboat hook and the davit

Figure 9.14   Partially enclosed lifeboat with fabric covers deployed

falls before actual release can take place (i.e. occurs when the lifeboat is fully waterborne).

On load release

This is defined as a system which is characteristically the same as the off load system but which can additionally be released, if required, while weight remains on the lifeboat hook and the davit falls.

NB. The hydrostatic safety device ensures that the release cannot be operated until the boat is waterborne. An overriding manual release operation is included to release the hydrostatic interlock in the event of an emergency.

Maintenance

It is a requirement of the Merchant Shipping Act that lifeboats and rescue boats are turned out and lowered under full safe working load conditions at least once every 5 years (an SWL loading of 110 per cent is the current requirement). When boats have been tested in this manner, or after being launched in normal use, when being hoisted and secured back in the davit holding position it must be ensured that, the hooks are closed and the release lever is in the locked position with the pin inserted. Once the hooks have been reset (most hooks are on a swivel arrangement), the crew at the bow and stern positions should check manually that the hooks are locked and in the closed position.

Several systems are used to restrain the hooks in position using 'half moon' cams or a 'pawl' arrangement connected to an operating cable (see Figure 9.15). Once activated the cam is turned effecting release.

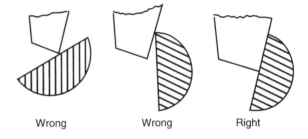

| Wrong | Wrong | Right |

Figure 9.15

Masters should ensure that Chief Officers make regular checks on release gear 'cams' and other operational features of the mechanisms, checking clearance levels and any effects of corrosion in particular, under the ship's planned maintenance schedule.

NB. A warning is issued by manufacturers that any work which has to be carried out on the boat release system *must* be carried out with the boat free of the davit falls. Do not attempt any work or physical checks, other than visual checks, whilst the boat is attached to the davit falls.

Hanging-off (preventer) pennants should be used to hold the boat or, alternatively, the boat can be lowered to the water (or the quayside) where the weight is taken off the falls to allow essential maintenance (see Figure 9.16).

Figures 9.17 and 9.18 show diagrammatically lifeboat release operation and the mechanism involved.

Free fall lifeboats

The design of vessels fitted with freefall launching systems (see Figure 9.19) is such that the boat which

Figure 9.16   Example of a single wire fall *in situ* on the boat's hook and the hanging-off pennant shackled into position

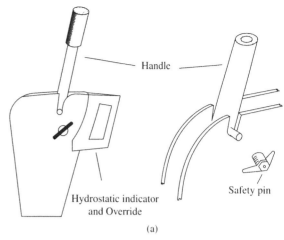

Figure 9.18   Lifeboat release mechanism

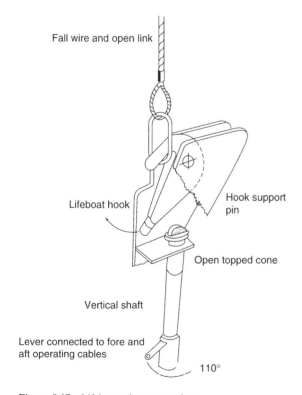

Figure 9.17   Lifeboat release operation

Figure 9.19   LPG vessel with all aft accommodation and freefall lifeboat *in situ* aft

accommodates the full ship's complement is launched over the stern of the vessel. Access into the craft is by a single stern hatch entrance which could, through lack of available deck space, limit the speed of entry. (crowd control and orderly boarding need to be). This coupled with the time to settle occupants and secure the body/head harness could make the procedure for leaving the vessel a time-consuming operation, up to releasing the craft.

Masters are advised that the idea of freefall has distinct advantages over and above the conventional lowering of survival craft and it is anticipated that more and more vessels will be built with freefall systems. The obvious advantages tend to outweigh the disadvantages but it is in the Master's interests to be aware of the time factor and have the boat ready and in a condition to launch with its full complement. Training drills with regard to cutting down the manning time of the craft could reap dividends in a real-time emergency.

Operation

The concept of freefall lifeboats has now been extensively employed in the marine industry on both merchant vessels and offshore installations. Different designs of survival craft allow crew to board at the sides or through the stern of the craft. Once on board, crew members would be secured in angled and well-padded seats by a body harness together with headbands to give full protection when launching.

Figures 9.20 to 9.25 illustrate the method of launch, accommodation and launching options for freefall lifeboats.

Figure 9.21   Helmsman's conning view of freefall lifeboat

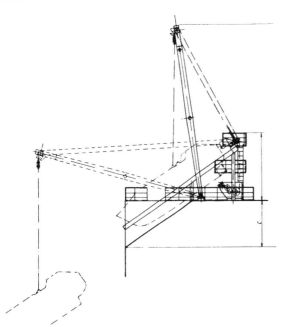

Figure 9.20   Method for launching and recovery of a freefall lifeboat

Freefall lifeboats initially built for oil rig platforms and tankers can now be seen on most types of vessels. Where they are carried under British regulations it is a requirement that an additional 'rescue boat' is carried, but this is not the case with all authorities.

The IMO requirements for freefall operation are such that the boat must be able to be launched against a 75° list and a trim of 20°. The height from which launching can take place varies from manufacturer to manufacturer but a launch from up to 40 metres would not be considered unusual.

The method of launching will differ between boat types but several effect release by use of the helmsman operating a hydraulic pump system. Alternatives to

Figure 9.22   View of the aft access point into freefall lifeboat

freefall include a hydrostatic release which allows the boat to float off if the mother ship is slipping below the surface, and a davit or crane system which permits lowering and recovery for drills and exercises.

The capacity, features and equipment of freefall boats are generally in keeping with other styles of survival

Figure 9.23 Internal view of freefall lifeboat looking aft

Figure 9.24 Internal view of freefall lifeboat looking forward

systems, tanker lifeboats being fitted with waterspray and self-contained air support for a minimum 10-minute period. Specialized features may include such items as self-bailing, protected/ducted propellers, internal lighting, searchlight, distress transmitter, self-righting ability after capsize, water-cooled engine, towing bollard, specific boarding arrangements from the water, raised

conning position, small gear lockers and under floor tanks, and shock absorbent seating with full body harness securing for occupants.

Figures 9.26 and 9.27 show freefall lifeboats stored in different positions.

Recovery operations

High sided vessels

Whenever Masters find themselves involved in rescue operations they tend to have to employ the resources available. At times the vessel may be fully loaded and at her subsequent load draught or in ballast with increased freeboard, situations that could hinder a recovery operation.

Ideally the launch of the rescue boat would be the obvious answer for the majority of cases depending on sea conditions and the general state of the weather. However, if other factors are considered such as the number of persons being recovered, and the medical condition of such persons, then the situation might generate the need for the use of alternative methods.

If the situation occurred where a Master of a large bulk carrier, in ballast, encounters a distress/recovery situation of, say, 12 persons of various ages, made up of four persons physically fit, four walking wounded and four stretcher cases, how could a rescue be effected in bad weather?

Without lifting gear on board the recovery vessel, the recovery of the survival craft would not be a feasible option. Cranes or derricks, provided they are of an adequate safe working load, could be used to hoist the survival craft with survivors onto the recovery vessel. The disadvantages of such an activity are governed by the type of survival craft to be lifted. Liferafts could be netted from underside and probably would not pose a weight restriction, whereas lifeboats could be expected to be in the region of 4 to 5 tonnes minimum, and would certainly be outside the limits for the smaller stores crane/derrick operation.

Topping a derrick or crane, even if the weight factor was not a problem, could very well present control problems especially in bad weather/high wind conditions and Masters would need to be mindful of the potential accident risk to their own crew. The complete hoist is without doubt an attractive proposition, provided weather and working loads on equipment permit, but there are too many 'ifs' associated with this option.

In the event of an incident some casualties may require a degree of 'self-help' if the use of embarkation boarding ladders, accommodation ladders, or even scrambling nets are to be considered. With stretcher cases, the Master needs to rethink how such cases could be effectively recovered from the surface. Casualties who

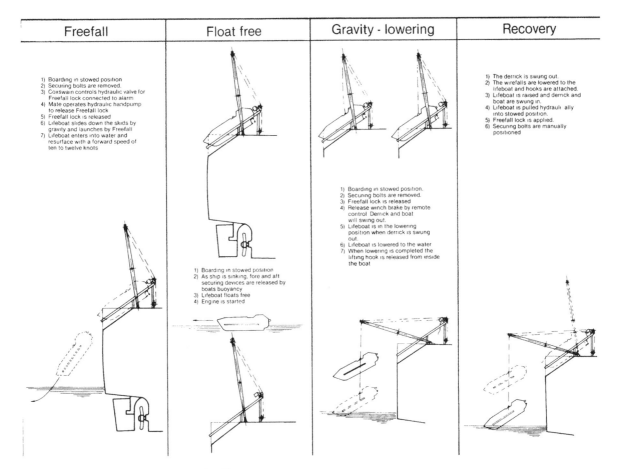

Freefall	Float free	Gravity - lowering	Recovery
1) Boarding in stowed position 2) Securing bolts are removed. 3) Coxswain controls hydraulic valve for Freefall lock connected to alarm 4) Mate operates hydraulic handpump to release Freefall lock 5) Freefall lock is released 6) Lifeboat slides down the skids by gravity and launches by Freefall 7) Lifeboat enters into water and resurface with a forward speed of ten to twelve knots	1) Boarding in stowed position 2) As ship is sinking, fore and aft securing devices are released by boats buoyancy 3) Lifeboat floats free 4) Engine is started	1) Boarding in stowed position. 2) Securing bolts are removed. 3) Freefall lock is released 4) Release winch brake by remote control. Derrick and boat will swing out. 5) Lifeboat is in the lowering position when derrick is swung out. 6) Lifeboat is lowered to the water 7) When lowering is completed the lifting hook is released from inside the boat	1) The derrick is swung out. 2) The wirefalls are lowered to the lifeboat and hooks are attached. 3) Lifeboat is raised and derrick and boat are swung in. 4) Lifeboat is pulled hydraulically into stowed position. 5) Freefall lock is applied. 6) Securing bolts are manually positioned

Figure 9.25 Launching options for freefall lifeboats

are unconscious, or suffering from severe burns, broken limbs or other similar injuries present a demanding situation when coupled with bad weather conditions.

The high sided vessel in Figure 9.28 presents a further problem when recovering persons from the water. To combat this certain high sided vessels may be fitted with specialized equipment which could be deployed to effectively recover persons from surface level, e.g. pilot hoists, shell (pilot/stores) doors, stern ramps of ro-ro ships, MES systems adapted with boarding ladders to allow recovery as opposed to disembarkation.

The problem of moving incapacitated persons from surface level to the embarkation deck of an attending vessel, has proved difficult to resolve. One solution is a rescue vessel employing its own lifeboats as an 'elevator' and would satisfy any type of vessel equipped with its own lifeboats and powered davit winch recovery systems. However, even this would involve a level of risk in carrying out a transfer from one survival craft to another, even though the rescue vessel is retaining its own lifeboat in the secured position on the falls.

Figure 9.26 Freefall lifeboat in the stowed position

If boats on fall wires are to be used as a transfer vehicle, then the lifting hooks should be moused and secured against accidental release. The manning of such an operation to effect the surface transfer might be more appropriate with the co-operation of volunteer seamen

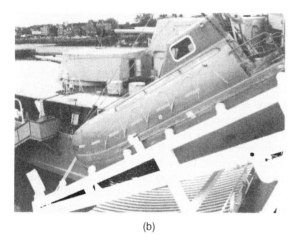

(b)

Figure 9.27   Freefall lifeboat secured in aft most position

Figure 9.28   High sided roll on-roll off ferry. Similar large ore carriers, or tankers when light or in ballast, can often present great difficulties for Masters if called upon to recover persons from the water

rather than ordering men into what could become a dangerous exercise. General advice prior to commencing a boat to boat transfer might also include instructions to secure both craft together to present a more steady, single unit. Where raft to boat transfer is being considered, the shape and flexibility of the liferaft may not suit being secured in case the raft drifted into the underside of the lifeboat and became trapped by the rise and fall of any swell.

The above system could be a useful method with a mother ship so equipped with lifeboats on falls positioned to port or starboard, but many vessels are now being fitted with freefall lifeboats with overstern launching positions. Such high sided vessels are not easily manoeuvred in high winds and to attempt recovery in the stern area by the above-mentioned method could compromise the use of engines and steering operations.

In any event the Master's ship handling capabilities could be called on in earnest to create a 'lee' for recovery from the surface. Station holding, especially without

bow/stern thrust units, and with or without controllable pitch propeller(s), could become extremely difficult.

Ship's features can be either beneficial or detrimental to an operation of this nature and an example of each is given below.

Roll on-roll off ferries

Ro-ro vessels have in the past used the stern ramp to launch rescue craft. This was possible because the hinge position of the ramp had an approximate freeboard of one metre and the ship could be held 'head to wind' with bow thrusters. It is debatable in post-*Esconia* days whether Masters would consider losing watertight integrity and risk opening a stern door while at sea on the assumption that hydraulics will remain active to reseal the ramp once the recovery is complete.

Roll on-roll off ferries are fitted with a lower 'belting arrangement' attached to the hull, just above the water line. If swell conditions were 'steep' a real danger of the survivors in their craft being trapped under this belting could become a dangerous reality.

Under the new changes to the SOLAS convention, roll on-roll off vessels are required to carry a 'fast rescue craft', but the same convention fails to be descriptive on what constitutes such a craft. The boat would be required to make 8 knots and carry at least six persons to meet specific criteria. It would also have to have an effective launch and recovery method for operation in poor weather conditions.

Such a craft could expect to be called on to operate in high seas and would be expected to demonstrate stability, endurance, as well as special features such as self-righting capability, to make it attractive to commercial operators.

Figure 9.29 Rescue boat situated on the starboard side of a roll on-roll off passenger ferry, stowed on single arm davit with quick single action release bridle

Figure 9.29 shows a starboard side mounted rescue boat fitted to a roll on-roll off passenger ferry.

Improvised alternative

The use of rescue/recovery/transfer baskets (see Figure 9.30) has been widely employed in the offshore oil/gas industry, where the installation's cranes are able to transfer from surface vessels to the higher rig levels. Such a method could with some improvisation be used in an emergency on board certain types of vessels.

Special designs of nets/baskets are also suitable for use from the underside of helicopters but it is unlikely that the majority of ships would carry a purpose-built equipment item of this nature. However, with a little thought and ingenuity it would not be inconceivable to construct a similar item from a cargo pallet and rope net. It would still necessitate the use of ship's lifting gear as, say, in the case of a derrick or crane, which may prove difficult to rig in heavy weather, although the risk element compared to employing the crew overside in adverse conditions must be considerably reduced.

The additional use of steadying lines and/or lifebelts attached to aid flotation at the surface could also be considered.

Figure 9.30 Capacity of purpose-built nets/baskets vary but a maximum load is usually no more than three persons. To use this type of equipment casualties need some self-help capability

Coastal and deep sea tanker vessels

Loaded tanker vessels with an already low freeboard may have the ability to ballast down even further, to effect a near float-on capability for would-be survival craft carrying survivors (on the principle of float on-float off heavy lift ships). However, the larger vessel in a light condition which finds itself in bad weather is extremely limited in options. Every Master has a responsibility to the crew and their welfare and would never willingly put them at risk. It must also be considered what the maximum level of weather would be before launching a rescue boat. Following many discussions with mariners on this point, the general consensus would be up to force '6' on the Beaufort scale. This level is by no means a yardstick, and would differ considerably for changing conditions.

The use of one of the recovery ship's lifeboats, set in conventional davits where the boat is lowered but not released from the falls, may offer a method of recovery for persons at surface level. This method uses the lifeboat as an elevator, and can be brought up and down the ship's sides to accommodate would-be survivors.

NB. This method would not be feasible if the recovering vessel was equipped with a freefall lifeboat, although such a ship might be able to use the recovery method for the freefall boat to effect survivor rescue in another form.

Recovery alternatives

When faced with a rescue operation Masters must consider the safety of the crew as much as the safety of the persons in distress. The situation may warrant *no* immediate attempt being made to effect recovery on the basis that an increased risk to life would be evident by exposing greater numbers of persons. Bearing this in mind it may be prudent to put a medic on board the survival craft to treat casualties and hold the situation until the weather conditions improve before attempting recovery.

NB. An alternative might be to throw over an own ship's liferaft to hold the survivor until the weather abates.

The geography of an incident could well dictate the use of helicopter recovery by the usual hoist methods of strop, basket or stretcher. However, helicopter use is strictly limited by range and endurance of the aircraft. In deep seas, a Master would have to decide when to move and how, and without doubt the state of the prevailing weather will be the influencing factor.

In any event, any rescue/recovery or holding operation should make full use of immersions suits, lifejackets, lifelines and harnesses, hard hats and foul weather clothing to reduce the overall risks.

Figure 9.31 shows a method of recovery using a fast rescue craft without a recovery aid and Figure 9.32 illustrates the use of the 'House recovery net'.

Figure 9.31 A fast rescue craft and crew demonstrate survivor recovery and the difficulties of rescue without a recovery aid

The role of the helicopter

The expanding use of helicopters in the marine environment is now established. The fact that more and more ship's personnel are experiencing routine as well as emergency ambulance service by helicopters, dictates the need for mariners to be far more aware of their own responsibilities when engaging with rotary winged aircraft.

Offshore development has evolved alongside helicopter transportation and the number of offshore vessels fitted with heli-deck landing areas continues to increase. Alongside this, seafaring personnel have had to move with the times regarding helicopter types, not only their recognition, but also their logistical needs. Endurance of the helicopter is based on fuel capacity and weight factor

Figure 9.32 Open boat, lowered to surface level, alongside mother ship on the boat's skates to rescue a survivor employing the 'house recovery net'. Boat's falls remain attached and crew have the benefit of lifelines

within its operational range. The aircraft does have its limitations, and cannot be expected to solve all problems all the time.

Masters must be conscious that during a helicopter engagement, their responsibility remains to the safety of the vessel, whereas the pilot's responsibility is the safety of his aircraft. To ensure a smooth trouble-free routine operation, the prudent Master would conduct an operational meeting of key personnel prior to commencing any helicopter engagement. Personnel who would need to be kept closely informed of the activity programme include Chief Engineer, bridge Watch Officer, helicopter (deck) landing officer, ship's Safety Officer, Bosun, Communications Officer (if carried), and the Chief Officer.

The degree of involvement of these and other persons aboard the vessel would largely depend on the nature of the operation, i.e. whether it is a 'hoist operation' or a 'land-on' situation.

The position of rendezvous between the ship and aircraft needs to be established well in advance and the Master is expected to ascertain that such an area was suitable and safe to conduct the operation.

Navigational requirements

The area of engagement is to have adequate underkeel clearance and sea room to be able to complete the operation. The Master should 'con' the vessel with full bridge team involvement. Communications should be established with the aircraft as soon as practicable and any data regarding surface conditions, weather, or ship identification should be passed.

The ship is expected to set a course, probably taking the wind direction into account, and the operational speed should be clarified prior to engagement. Correct navigational signals should be displayed on approach and throughout the period of engagement.

Table 9.1 shows the use, range and capacities, etc. of various helicopter types.

Vessel identification to the helicopter

Where only a single ship is inside the area of engagement it is not difficult for the pilot of the aircraft to identify the vessel; however, where several ships are in the vicinity the Master should ensure that everything to distinguish the ship as being the rendezvous vessel must be made.

1. Communication of the ship's name and the type of vessel, i.e. tanker, container vessel, roll on-roll off ferry, etc.
2. Position of ship and ETA at the position of engagement together with the inbound course direction.
3. Radio homing signal transmitted from the vessel.

4. Description of the paintwork, colour scheme of the ship's superstructure.
5. Description of ship's design and any special features, e.g. all aft accommodation. Badge or emblem on funnel.
6. Signals exhibited: ball diamond ball (red white red at night), and any international code flags being shown.
7. Deck markings: 'H' or 'Spot' marking of the deck operational area.
8. Nationality of the ship and position of its ensign.

The Master is expected to check and confirm his activities to ensure that they conform to the current *ICS Guide to Helicopter Shipboard Operations*.

Emergency ambulance use of helicopters

The use of the helicopter in a marine emergency to evacuate a single casualty or to extract several persons at

Table 9.1

Helio – Type	Employment	Capacity	Ops/Range	Max. Range	Remarks
Dauphin 365	Offshore ferry	12	253	506	All purpose.
Super Puma 332L Ferry	O/S ferry, SAR	24	340	1080	Civilian and military
			Aux. tanks	Ferry tanks	
Bell 214 ST	O/S transport	18	250	505	External lift capability
Westland 30	O/S passenger	19	185	481	Medivac alt. config.
				Aux. tanks	
MBB Bk 117	Inter-rig transport	8/12	135	500	Utility aircraft O/S
			Stand tanks	Aux. tanks	
Sikorsky S61N	O/S ferry and SAR	24	215	495	Amphibious O/S
				30′ reserve	
				transport	
Sea King	Military – SAR	22 max	250	664	5.5 hrs endurance military options ASW
Sikorsky 76	Offshore duty	12	200	600	All-purpose O/S
				Reduced	
				load + aux. tanks	
Bell 412 SP	O/S Transport	14	250	400	Ferry operations
EH101 Merlin	Military and civilian	30	275	500+	SAR and utility, ASW O/S and military ferry
Seahawk SH-60B	Military USN	10	200	400	SAR, ASW, medivac
Sikorsky Jayhawk medium range recovery	USCG	4 crew + 6	300	plus 45′ on scene	Multi-mission with extended range
Sikorsky Sea Dragon Super Stallion CH-53E	USN	55	1000	in flight refueling unlimited	Troop transport, mine clearance
Puma 330J	O/S utility	19	up to 500	Extended Range Tanks option.	All-purpose utility
Chinook	Military and Civilian	44	575	with long range tks 840	Long range ferry, tandem rotors, heavy lift capability

once in heavy seas is probably the most ideal of vehicles. Search and rescue (SAR) is no longer just the domain of military aircraft although they still tend to dominate the activity and make a tremendous contribution in any disaster situation. Civilian authorities, such as Coastguard Agencies, large oil companies, offshore operators, local councils as well as independent agents, are actively engaged in SAR activity. Standards throughout the aviation industry are by the very nature of the business extremely high but operational practice can expect to vary between countries and operators.

It is unlikely that Masters and their crews who find themselves on the receiving end of emergency helicopter assistance will have much time to plan anything and the operation could well be taken out of their hands completely. Once the helicopter(s) is/are on site, they will immediately carry out pre-hoist checks and assess the capability and the risk factor. Masters will probably not be aware of exact payload capacity, endurance of the aircraft and other essential operational details and as such would need to accept from the pilot of the aircraft instructions as to the best course of action.

Foreknowledge that the aircraft involved is not without limitations, e.g. operating range, endurance, lifting capability, payload capacity for walking wounded and stretcher cases, is going to directly influence just what the aircraft can and cannot achieve. The time of day and weather conditions also influence operations in certain cases, i.e. fog conditions could keep aircraft grounded, while storm force winds could prevent lift off, unless the aircraft are located with an internal hanger lift-off facility.

NB. Night flying capability requires aircraft and aircrew to be suitably operational and passed to Instrument Flying Rating (IFR) standards.

Being forewarned is advantageous in a hazardous situation. Panic and fear amongst survivors could set in if a rescue vehicle suddenly broke off contact and departed the scene without a seemingly credible reason. If these fears can be allayed with a logical reason, morale might not suffer as much.

The Hi-Line operation

Helicopter operations and the type of engagement will obviously vary upon the circumstances at the time, but many authorities are more and more favouring the use of the 'Hi-Line' method as opposed to the single vertical hoist. The Hi-Line is to be employed when:

(a) there are a lot of casualties to transfer,
(b) the weather is extremely bad,

Figure 9.33 An S61N helicopter, (civilian variant of the Sikorsky 'Sea King') operating with HM Coastguard, engages in hoist/training activity with a small deck pleasure craft. NB. The speed of the surface craft is high but can be easily matched by the aircraft

(c) there are surface obstructions like masts and aerials which could foul a vertical hoist operation.

Principle
The Hi-Line method allows the aircraft to hold station alongside the surface vessel, in full view of the pilot situated on the starboard side of the helicopter. The movement of casualties between the two vehicles alongside each other is the key operational feature unlike the scenario where the aircraft holds position directly above the casualty prior to making a vertical hoist, which often leaves the pilot 'blind underside' and relying solely on verbal reports from the observer. Double hoists are made once the station holding and the Hi-Line is established between both the helicopter and the surface craft. Both vehicles remain on parallel courses while the persons being recovered move between them.

Operation
Once the aircraft has established a rendezvous with the surface craft and completed all pre-hoist checks, a weighted heaving line will be thrown from the access of the aircraft to the deck of the ship. Alternatively the line could be trailed towards the vessel as the aircraft traverses sideways over the surface target, before returning to a station holding position on the ship's beam.

Once the heaving line is held by the deck crew, it is normal practice for the hoist wire and double strop arrangement to be lowered. In the first instance an aircrew/frogman is normally landed on the first hoist, with the intention of controlling the operation at surface level. The idea is for the deck crew to heave in on the vertical hoist by means of the Hi-Line, and so effectively pull the person in the strop to land on the deck. Once landed he would control the despatch of casualties/survivors by means of the double strop arrangement into the aircraft.

NB. The aircrew/frogman when lowered would normally be holding the Hi-Line, and so by-pass a weak link arrangement set into the line. As the deck crew heaved in on the line, the person inside the strop would not have their arm pulled off, but be drawn into a landing on the deck. (The purpose of the weak link is to ensure that the aircraft can always break off the engagement even if the line became snagged.)

Sikorsky HH-60J Jayhawk medium range recovery (MRR) helicopter

Aircraft information
The Jayhawk first became operational in 1991 as a medium range recovery helicopter. The initial role was that of search and rescue (SAR) together with a law enforcement role by way of drug interdiction. In its primary role of SAR the aircraft can operate up to 300 nautical miles offshore and maintain an on-scene endurance of 45 minutes. When so engaged and operating with a crew of four, the aircraft can transport six survivors in a walking wounded capacity.

The performance of the aircraft is enhanced by a requirement for it to be able to operate in storm force '11' winds at up to 63 knots and in exceptionally heavy sea conditions. It is fitted with an RDR 1300 search weather radar, advanced avionics and communication facilities. It can also operate an extended range option by incorporating 3 × 120 gallon external pylon drop tanks.

Mission systems include:

- UHF/VHF/FM radio, HF radio, and crypto computers, radar altimeter and GPS provisions in addition to the search weather radar.
- Dual redundant mission computers and cockpit video display.
- 600 lb rescue utility hoist and a 6000 lb external hook capacity.
- Triple redundant hydraulic and electrical systems within the crashworthy airframe.

Jayhawk helicopters are in operation on both the eastern and western seaboards of the United States, engaged with the US Coastguard. They are also in service with the Coastguard Training Center in Mobile, Alaska, and are compatible for operations with the 'Hamilton' and 'Bear' class Coastguard cutters.

Multi-mission roles
The aircraft continues to be involved in marine environmental protection and pollution control, aids to navigation, military readiness and support, logistic support, drug traffic prevention and other similar law enforcement activity.

The Jayhawk is scheduled to replace the Sikorsky HH-3F Pelican series and provide added support to the US Navy in maintaining the Maritime Defence Zone by way of anti-submarine patrols, mine countermeasure operations and combat search and rescue activity.

Figure 9.34   A US Coastguard Sikorsky Jayhawk

Helicopter evacuation (medical) check off list

When communicating to the Marine Rescue Co-ordination Centre (MRCC) or to the Rescue Co-ordination Centre (RCC) for medical evacuation the following should be provided:

General information

1. Vessel's position and name.
2. Vessel's course and speed.
3. The next port of call with date and ETA.
4. Type of vessel, i.e. container, tanker, supply, etc.

Patient's condition

1. Possible diagnosis, if known.
2. Current condition of patient.

3. Any medication that is currently being administered to the patient.
4. Pulse.
5. Blood pressure.
6. Previous medical history of patient.
7. Any medication that the patient is allergic to.
8. Is the patient ambulatory or incapacitated.

Weather Conditions, current

1. Sky/cloud condition.
2. Estimated cloud ceiling.
3. Precipitation in sight.
4. Wind speed and direction.
5. Sea state conditions

NB. When carrying out evacuation of a patient relevant documentation should be dispatched and landed at the same time, e.g. passport, medical report, list of medications and dosage administered.

Post rescue collapse and influencing factors affecting helicopter SAR activity

With the many rescue operations conducted with helicopter participation it is only to be expected that aircrew have gained vast experience from the successes and from the failures. Extensive investigation has taken place by the Institute of Naval Medicine regarding the recovery of hypothermic casualties, which revealed factors relevant to helicopter recovery operations.

In the case of a survivor who is immersed in water it was found that the body experienced a level of water pressure known as 'hydrostatic squeeze'. However, when the casualty was lifted clear of the surface, the body suffered a loss of this pressure. During or after hoisting towards the aircraft, the heart rate of the casualty showed a distinct increase. This increase was to compensate for reduced venous pressure and also due to a normal reflex action. When coupled with considerable physical exertion following removal from the water, would-be survivors could and did suffer from 'post rescue collapse'.

It was noted that three main factors affected post rescue collapse:

1. Loss of hydrostatic squeeze.
2. The adverse gravitational effect on blood circulation that results from a vertical lifting operation.
3. Physical exertions of the casualty during the rescuing phase of the operation.

In order to rescue and treat the casualty, removal from the water must take place, and for obvious reasons, the sooner the better. So it is inevitable that loss of hydrostatic squeeze will take place. However, the method of hoist and the physical exertions placed on the casualty could be changed in an attempt to minimize adverse effects.

Lifting methods which employ a double strop for a single casualty have already been tested and observations showed little change on the heart rate of a person being hoisted in a near horizontal position.

NB. The same person being hoisted by the vertical, single strop, standard hoist experienced a 35 per cent increase in heart rate.

It would therefore seem logical that the horizontal method of hoisting is preferred to the conventional vertical lift. Aircrew experienced some difficulty in the use of the double-strop alternative, especially in bad weather of more than force 4, where wave height could be above 1.5 m.

Current practice continues to employ the conventional vertical lift for speed of operation, but more operators are now employing double-strop recovery methods where the circumstances dictate. The basic idea of speed of recovery has not been lost and still remains an essential ingredient to combat the effects of immersion and exposure.

Merchant ship search and rescue operations

For additional reference see the new IAMSAR manuals.

> The Master of a ship at sea, on receiving a signal from any source that a ship or aircraft or survival craft thereof is in distress, is bound to proceed with all speed to the assistance of the persons in distress informing them if possible that he is doing so.
> *Regulation V/10 of the International Convention for the Safety of Life at Sea (SOLAS) 1974*

The obligation placed on the Master of a merchant vessel to respond to a distress situation will probably bring with it the greatest responsibility of any situation which might be encountered in the marine environment. One of the purposes of the IAMSAR manual is to provide guidance during an emergency involving search and rescue operations. Few Masters would be knowledgeable of all their obligations and duties regarding specific roles and reference to the IAMSAR manual may be reassuring.

Acknowledging the distress signal, and then responding to provide assistance, is an act of responsibility in itself but the level of responsibility could escalate considerably, especially if the responding vessel is the first vessel at the incident and the ship is charged with the function of On-Scene Co-ordinator (OSC).

The task of a rescue unit, or that of an On-Scene Co-ordinator (OSC), for example, is not considered an every day activity at sea and a Master would not wish to be found wanting when circumstances have generated an emergency. It is what every seaman has trained years for, but hopes never to put into practice. When an incident does happen; the circumstances are never as one imagines them to be.

A vessel could be working in conjunction with others or in isolation. Either way, the need to exercise 'power of command' will never be greater than when faced with a distress scenario. The Master will be called on to command and probably make life threatening decisions – hopefully these will be the correct decisions.

For additional reference see the emergencies and search and rescue procedures found in Chapters 6, 7, and 8 of *Seamanship Techniques*, Part 2.

The IAMSAR manuals

Purpose

The primary purpose of the three volumes of IAMSAR is to assist states in meeting their SAR obligations agreed under the convention on International Civil Aviation, the International Convention of Maritime Search and Rescue and the SOLAS convention.

The volumes are designed to provide guidelines for a common approach by the maritime and aviation industries towards the organization and provision of SAR operations.

Volume I

'The Organisation and Management' for global SAR systems. This volume presents an overview of the SAR concept and introduces key components of the system, including communications, support facilities, rescue co-ordination centres, and the function of the On-Scene Co-ordinator. Consideration is given to the desired training and qualification of SAR personnel and the effective co-ordination of the system components following an alert. It is also directly concerned with the management perspective of an SAR system and recommends some techniques which could expect to encourage improvements within the services.

Volume II

'The Mission Co-ordination' to assist personnel who plan and co-ordinate SAR operations and exercises. This volume presents an overview of the SAR concept from a global view as well as a regional and national view. It is directly concerned with the phases of an alert, from awareness through the distress period, to the recovery phase. Search planning and search techniques of small and large areas are considered with the limited available facilities to provide a viable search action plan. The volume also provides guidance regarding modes of rescue, completion or suspension of search and the completing of final reports.

Volume III

'The Mobile Facilities' intended to be carried by rescue units, aircraft and vessels to enhance the performance of search, rescue or On-Scene Co-ordinator. This volume is concerned with the practical activities for on-board emergencies, working with helicopters, recovering and treating survivors, etc. Communications and distress signals together with effective response are detailed in this particular publication. One which Masters and Watch Officers will find extremely useful in the event of an emergency incident.

Master's orders on route to an SAR incident

Every SAR incident will be different in the type of craft involved or the number of survivors/casualties that are affected. However, standard procedures are expected to cover most eventualities, but on occasion additional resources for large numbers or a specific ship type may necessitate specific actions to satisfy the current situation. In general a Master proceeding towards an incident, assuming the role of a rescue/recovery unit, would make the following preparations:

1. Establish a bridge team in situation and place the engine room on immediate standby (Master taking the 'con' of the vessel and switching to manual steering).
2. Instruct the Chief Officer to prepare the deck to receive survivors/casualties:
   (a) turn out and make the rescue boat ready for launch,
   (b) ensure rescue boat crews are correctly attired and briefed,
   (c) turn out and rig the accommodation ladder, ready for lowering,
   (d) secure a guest warp on the anticipated 'lee side',
   (e) have scrambling nets and/or pilot ladder facilities ready for use.
3. Order the Officer of the Watch to display appropriate signals in compliance with the International Code of Signals.
4. Order the Medical Officer to prepare the hospital and other accommodation to receive casualties. Place the first aid party at a state of readiness to treat for specific injuries.

5. Order the Communications Officer to establish and retain contact with the On-Scene Co-ordinator (OSC) and/or the MRCC. Obtain an updated weather forecast for the operational area. (Communications are expected to be a continuous activity and will include a variety of specific signals between various parties including the casualty, other search units, the OSC and MRCC.)
6. Order the Navigation Officer to chart the search area and plot a designated search pattern to include the commencement of search pattern (CSP) position.
7. Brief personnel on the incident and advise lookouts on possible target definition.
8. Order catering personnel to prepare hot food for survivors/casualties.
9. Continue to maintain an effective watch culture, by way of lookout, position and underkeel clearance.
10. Update all parties by both internal and external communications on the progress of the incident.

The duties of the On-Scene Co-ordinator (OSC)

The IAMSAR manual states that the person in charge of the first SAR resource to arrive at the scene will normally assume the function of OSC, until the SAR Mission Co-ordinator (SMC) at the Rescue Co-ordination Centre (RCC) directs that the duty is transferred (presumably to a more suitable vessel, e.g. a warship).

It is, however, recommended that the designation of an On-Scene Co-ordinator (OSC) is appointed as early as possible and preferably before arrival within the specified search area. In practical terms an OSC would be best suited to a warship for several reasons, not least that a man of war would have superior communications and plotting resources as well as available manpower. Neither would it be compromised with hazardous or perishable cargoes. It may also carry its own reconnaissance aircraft and probably have a longer endurance period than a commercial vessel.

It is not inconceivable that the OSC would have to conduct similar duties as the SAR Mission Co-ordinator (SMC) and plan the actual search if that vessel became aware of a distress situation and was awaiting to establish communications with an RCC. As such the following activities could fall within the OSC role:

1. Allocating search areas and designating appropriate search units.
2. Allocate working radio channels to specific units.
3. Advise on search patterns and recommended track space.
4. Collate rescue information and resources available from other search units.

5. Communicate updated weather reports to concerned parties.
6. Obtain progress reports from search units and relay to MRCC.
7. Request additional facilities/advice from MRCC as required, e.g. helicopter use, or air droppable equipment.
8. Update all parties following survivor debriefs or on changing circumstances.
9. Monitor the endurance capabilities of respective search units.
10. Organize medical/evacuation facilities as appropriate.
11. Contact relevant ship reporting authorities, i.e. AMVER, AUSREP.
12. Inform search units of outcomes or when concluding the operation.

The designation of an On-Scene Co-ordinator is meant in part to mirror the activities of the shore-based Maritime Rescue Co-Ordination Centre (MRCC) by monitoring the progress of a number of search units and acting as a communication conduit between the search units and the MRCC

Any vessel acting as a search unit or as an OSC is expected to maintain a detailed log of operational detail. The importance of this, especially in the case of the OSC, cannot be overemphasized and would be considered essential evidence at any subsequent enquiry.

IAMSAR case study

Scenario–22nd February 1998

Aboard the MV *Probitas* steaming approximately 10 miles off the coastline of Morocco.

Time	Remarks
2306 LT	A frantic call received on VHF Channel 16 that the vessel MV *Adria* was on fire. (No MAYDAY was used and the type of assistance required was not specified). The Master was informed that the vessel on fire was only 12 miles from the position of *Probitas*. The Master immediately took the 'con' of the vessel and the Chief Officer took over communications. A general call to all shipping was made and it was realized no other vessel had received the presumed distress call.
2320	The MV *Adria* made a second communication that she was on fire

following an engine room explosion and that she had made a distress IMARSAT-C alert to Norway. (The transmission was nervous and poorly communicated.) Communications then ceased abruptly.

MV *Probitas* then made a second call to all shipping and it was confirmed no other vessels had heard the communications.

The Master of MV *Probitas* subsequently transmitted a MAYDAY RELAY and obtained a fix on the striken vessel.

The *Probitas* altered course towards the MV *Adria*, and the deck preparations were made to effect recovery of survivors.

2340 (approx.)	Other shipping started to contact MV *Probitas* to obtain updated information but did not respond until *Probitas* requested assistance from all area shipping.
0012 (23/02/98)	*Probitas* sighted MV *Adria* in flames and confirmed the distress call by a further MAYDAY RELAY.

The *Probitas* estimated that the MV *Adria* was a general cargo vessel of about 2000 MT, and that she was extensively on fire from the accommodation, bridge, and engine room. It was felt that because the fire was so intense that survivors would already have abandoned the vessel.

MV *Probitas* (tanker vessel) held station approximately within 1 mile of the stricken vessel (*Probitas* was in ballast and her tanks were not inerted or gas free).

Contact was made to RCC Stavanger (Norway) to provide on-scene detail.

0015 LT	SATCOM from RCC Stavanger requested MV *Probitas* to conduct the mission. The Master of *Probitas* agreed and the ship was designated as On-Scene Co-ordinator. (This designation was confirmed to all ships by RCC Stavanger requesting them to report to *Probitas*.)
0017 (approx.)	Seven ships responded to MV *Probitas*, one of which was not retained because of language difficulties.
0018	Lookouts aboard *Probitas* reported a distress rocket sighted and a flashing

white light was then seen 2–3 cables from the stricken vessel.

This was observed to be a survival craft and the signal was acknowledged by aldis lamp and ship's fog horn.

NB. The weather at this time was estimated as force 6/7, strong winds and heavy seas being experienced.

The position of the survival craft and the weather conditions were such that it would have been hazardous to attempt to launch the rescue boat from *Probitas* or to position *Probitas* too close to the vessel on fire.

MV *Probitas* subsequently ordered the MV *Rosa Tuckano*, a roll on-roll off vessel which had reported to assist, to attempt to recover survivors from the liferaft.

0056	*Rosa Tuckano* arrived on scene and it was decided that MV *Probitas* would create a 'lee' for the survival craft while the *Rosa Tuckano* used her thrusters to come alongside the liferaft to recover survivors.
0130	Nine survivors were recovered by the *Rosa Tuckano*.

Following a debrief of the survivors aboard the MV *Rosa Tuckano*, it was realized that the distressed vessel was Romanian with a crew of 16. Three people were reported dead following an engine room explosion and the remaining four crew members, including the Master, had sought shelter on the forecastle head of the MV *Adria*. (No survival craft was known to be situated on the fore deck of the distressed vessel.)

RCC Stavanger was updated about the nine survivors, who were generally well but with minor burns. MV *Probitas* requested helicopter assistance to rescue the four remaining men aboard the MV *Adria*.

RCC Stavanger later informed the *Probitas* that the MRCC Madrid had despatched a helicopter from Cadiz to assist. (Helicopter given an ETA of 1 hour.)

Communications had now been established between the four men on board the distress vessel and the *Rosa Tuckano*, which had taken up a position of about 100 metres away from the ship. The survivors were informed that help was on the way and because the cargo holds were intact and the vessel not listed, the decision to wait was taken. Searchlights of both vessels were kept trained on the distressed vessel.

0243 LT — An SAR helicopter arrived on scene but could not remain on station due to limited fuel on board. The aircraft departed to refuel at Casablanca.

The decision was then made to attempt a ship to ship rescue, using the MV *Rosa Tuckano* to manoeuvre in close to the distress vessel.

0330 (approx.) — Following consultation it was decided to make contact by using a rocket line, but this proved to be in vain. The initial idea being to pass over two lifejackets for the two survivors who were currently without any buoyancy aid.

The *Rosa Tuckano* was to manoeuvre to within 20 metres of the vessel on fire and was eventually able to pass a heaving line over. Messengers were then married together and the six person liferaft from the roll on-roll off vessel was inflated and floated down to the MV *Adria*.

0545 — The *Rosa Tuckano* had successfully recovered the four remaining survivors from the liferaft and the Master of the *Adria* confirmed the complement as no more than 16.

0630 — Assistance was no longer required and the distress was cancelled.

All participating vessels were ordered to stand down and resume normal operations.

10 Marine training

Introduction

The benefits of continuous on the job training have yet to be reaped. The current feeling towards marine personnel learning through the workplace is not new. It was conducted in the 1950s and 1960s with seagoing apprentice schemes but these methods left a necessity for lengthy study periods on shore for obtaining qualifications. The student experienced a loss of salary and the shipowner lost the benefit of the employee aboard the ship, not to mention having to meet large study leave costs.

The National Vocational Scheme (NVQ), when conducted and operated in a correct manner, provided the industry with the qualified officer while retaining him or her *in situ* aboard the vessel for all but a short period of time. The recording of tasks experienced by the individual inside a portfolio is a means of producing evidence to satisfy an examination body. For this portfolio to be acceptable, ship's officers charged with the task of on-site recording of seamanlike tasks, must participate diligently in the training of younger seafarers.

The credibility of any system will eventually depend on the people who operate the scheme and to say such a system is without fault would only disguise its shortcomings. Clearly, it must be assumed that the officers charged with testing and examining the tasks in the workplace are qualified in assessing others in specified areas of competence. To this end many persons ashore and afloat are being trained as 'internal assessors' with D32 and D33 qualifications.

This alone will not provide complete safeguards to a system which is open to abuse. There is no guarantee that an assessing officer is confident in the task being undertaken by the trainee. Neither is it acceptable to assume that all senior officers have the time, the patience or the will to become consciously involved. The role of the Master as a guide to direction in the training of junior officers must be realized as being paramount. Without such direction, the future of the industry could well be passed to unskilled personnel inside irrelevant organizations. This in itself could have long-term repercussions for the quality and ability for our seamen and our merchant fleets in the future. If the industry does not police

itself correctly then external bodies could and will point the finger of retribution when accidents occur.

The Mercantile Marine has often been referred to as the fourth arm of the military. Many campaigns involving Army, Navy and Air Force have relied on the logistical support provided by the Merchant Marine. All the military services consider themselves on a continuous training programme whether they are actively or passively engaged.

On-board training is the ideal, but certain activities cannot be fully catered for or totally controlled in a shipboard environment. To this end an amiable relationship with shoreside training establishments is essential. On-shore training can not only be compatible but provide an essential element that cannot be obtained afloat. A typical example of controlled, safe training for the offshore industry is shown Figure 10.1, where a ditched helicopter transport is simulated requiring emergency underwater escape methods. Such an activity would be costly to set up offshore and would be difficult to control, performed shoreside, such an activity can give trainees a 'virtual reality situation' without the associated risks of an uncontrolled environment.

The training manual

It is now a requirement under the SOLAS amendments that all vessels now carry a training manual for survival at sea. This manual, which can be in video format, should be placed in mess rooms accessible to crew members.

One of the objectives of the manual is to provide adequate foreknowledge to support the on-board training with the ship's safety appliances and in particular the life saving survival craft carried aboard the parent vessel. The manual is meant to enhance the boat, fire and emergency drills carried out aboard the ship not replace them.

Shoreside supportive training

There are of course many different manufacturers of survival equipment worldwide and it would be

Figure 10.1 Simulated helicopter capsize using the rotating 'dunker' in the controlled environment of a water training tank with diver/instructor support

Figure 10.2 Shoreside survival craft training platform, with totally enclosed lifeboat, survival systems international capsule, fast rescue craft and single arm davit displayed. This set up operates up to a 12 metre rise and fall in an exposed weather environment

unreasonable to expect every seafarer to be familiar with every item of equipment. The ship's equipment can be expected to remain reasonably static once installed aboard the vessel but the movement of labour from one ship to another is not fixed. Each seaman will learn from past experience and it is hoped that familiarity will breed confidence in many different varieties of marine equipment. Shoreside training establishments can provide supportive training in semi-controlled environments (see Figure 10.2).

Practical training with external agencies

The obvious benefits of combined training with external agencies cannot be emphasized too highly. Provided exercises are well planned and communications are good the outcomes can be beneficial to all parties. The military tend to welcome training alternatives and as they are often called to assist in civilian as well as military maritime emergencies the experiences gained are generally appreciated.

Familiarity with each other's procedures, terminology and equipment can be beneficial in real-time incidents. Problems can be addressed before being experienced in the field and practised measures tend to operate more smoothly (Figure 10.3).

Figures 10.4 to 10.9 show training exercises using a variety of equipment.

(a)

Figure 10.3 Royal Air Force Wessex 5 helicopter engages with a Nautical College training vessel and Merchant Navy personnel. Aircrew and college staff both gain from the experience, enabling them to experience each other's terminology and equipment

(b)

Figure 10.3 (*continued*)

Figure 10.4   Training with Coastguard officers in the use of rocket line throwing gear

Training incident report

In the early part of 1999, a training exercise to evacuate a vehicle ferry in UK waters was conducted. An MES

Figure 10.5   Simulated night training with the use of the helicopter strop over water

was deployed and the Coastguard helicopter, a Sikorsky S61N, participated in the evacuation activity.

The ferry evacuation exercise took place inside the confines of a harbour sheltered by a breakwater. The weather conditions at the time were a Beaufort wind force '5', with a slight sea running.

About 40 persons had descended the marine evacuation slide deployed on the port side, aft quarter and were aboard a liferaft as the helicopter flew at about the height of the upper deck from aft to forward. As the helicopter passed over the liferaft and the MES the combined weather conditions and the down draft from the helicopter's rotors caused the MES slide to 'Jack-knife'.

It was fortunate that no persons were on the slide at the time. However, observers noted that although nobody was injured and the slide returned to its normal position once the helicopter had cleared the area, casualties could have suffered crushing injuries if they

Figure 10.6 Combined exercise with fast rescue craft, RAF Wessex Mk 5 helicopter and coastal vessel, participating in hoist/recovery training

Figure 10.8 Rescue boat crews being instructed in stretcher/first aid practice with immersion victims

Figure 10.7 Demonstration of correct donning of immersion suits during a boat drill

had been engaged on the slide at the time the aircraft passed by.

Such training activities can generate positive feedback on procedures and equipment with clear lessons to be learned for participants. In this notable incident helicopter pilots and aircrew should be extremely wary about hovering in close proximity to a deployed marine evacuation system, because of the down draft from rotors.

Masters and ship's officers, if in communication with pilots, should advise of potential hazards and the dangers of drawing too close when an MES is deployed and in use.

Practical training: semi-controlled conditions

It has long been appreciated that training activities need to be safe as well as effective. To this end such training

Figure 10.9 Aviation heli-raft with canopy deployed being used in training exercise following simulated helicopter ditching incident

must be as realistic as possible without being foolhardy. This has never been more desirable as when dealing with the associated dangers when training with the raw elements of fire or cold.

Figure 10.10   Shoreside fire-fighting facility

Figure 10.11   Totally enclosed boat passes through burning surface oil to demonstrate the fire protection afforded to occupants

Shoreside facilities can provide the controlled training environment for fire-fighting which could clearly not be carried out as on the job training. In these facilities mariners can experience the problems of heat and smoke, tank fires and other similar enclosed space fire-fighting scenarios.

On board the vessel, attention to drills and the familiarity with the use of equipment can be combined with the shoreside experiences to provide a professional approach to the dangers of fire.

Fire protection afforded to the occupants of a totally enclosed boat is demonstrated in Figure 10.11.

STCW '95 certificate requirements

The new certificate requirements for Merchant Navy officers now includes additional training in survival and fire-fighting practice. The Certificate of Proficiency (CPSC) has been extended to include rescue boats (CPSC&RB) other than fast rescue boats (FRCs). Personnel who may be called upon to launch and take charge of an FRC as a coxswain must meet additional standards to demonstrate a level of competence in this specific field.

Advanced fire-fighting has become a requirement for personnel who are expected to control fire-fighting operations, ideally to reflect management skills when employing personnel and resources to combat fire incidents in the marine environment. Passenger ships and roll on-roll off vessels also expect their shipboard personnel to have crisis management skills and training in aspects of human behaviour/stress handling and communication ability.

Figure 10.12   Example of full bridge training simulator based at the Fleetwood Nautical Campus in Lancashire England

Medical care requirements have changed and have moved closer to the standards expressed by the International Labour Organization (ILO) responsible for the global labour standards which relate to seafarers' employment. Seafarers who are designated as first aiders must be trained to higher standards than the basic needs required by seafarers in general. Effectively the revised standards are directed towards persons who are appointed to take charge of medical care on board and

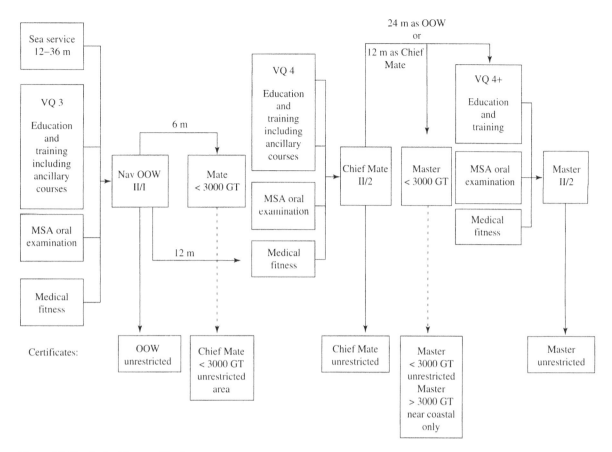

Figure 10.13 Desk officer certification

such personnel must produce documentary evidence or a certificate of having completed an appropriate training course.

The STCW standards are now established to ensure that all training and assessment of competence, certification and endorsement are meeting the requirements of a quality standards system. There is now a greater emphasis on the outcome of training and that such training provides the seafarer with the ability to perform his/her duties in a competent and safe manner.

In support of this, the revised convention has introduced mandatory requirements concerning the performance standards for simulators. Bridge watchkeepers will be required to demonstrate a level of competence with radar and ARPA equipment and general watchkeeping practice by means of 'ship simulators'.

Full mission simulation for marine students

The on-line exercise in Figure 10.12 shows a vessel entering Sidney Harbour. The control console and electronic navigation chart are centrally positioned. Collision avoidance and ARPA radar screens can be seen either side with ship control display screens providing continuous vessel status. The simulator is also equipped with manual or autosteering with single or twin engine/propeller options.

This example operates with up to five independent ship's bridges, each monitored by a single supervisor incorporating black box, visual and audio recording capabilities. Mission activities include mooring operations, ship handling, passage planning and shipboard operations. Software options allows a variety of geographic areas to be included permitting real-time pilotage and navigational training to all sections of the maritime community.

The international standards of training, certification and watchkeeping (STCW)

By now the majority of seafarers will have heard and are aware of the changes that are occurring in the developing standards worldwide, amongst mariners. The STCW convention was adopted by the International Maritime Organization (IMO) in 1978 and its underlying concept was to promote the 'safety of life and property at sea and protect the marine environment'. This aim was to be achieved by establishing international standards of training, certification and watchkeeping for seafarers.

At the time of writing some 96 countries had ratified the convention, amounting to approximately 96 per cent of the world's fleet. The regulations can also be brought into effect on non-compliance vessels as and when they visit ports of countries which have adopted the convention. Vessels which could endanger life or endanger the environment could be detained following an inspection through the Port State Control authority, until such time that the vessel is rectified and she no longer presents a problem.

STCW: basic function

The convention lays down basic training standards which participating countries are expected to meet or exceed in the areas of maintaining:

(a) a navigational watch,
(b) an engineering watch, or
(c) a radio watch.

Additionally, minimum requirements are specified for the training of mariners who are engaged on tanker vessels and roll on-roll off passenger ships, and for all seafarers to be effectively trained in the use and proficiency of radar, communications, survival craft, medical care and fire-fighting practice.

Officers: certification

Minimum requirements are stipulated for the certification of the ship's Master, officers and crew, with provision for updating and ensuring a continued level of proficiency.

Navigation Officers (STCW '95) will require one of the following:

- Watchkeeper's Certificate,
- Mate's Certificate,
- Master's Certificate,

for vessels engaged on international voyages.

NB. The regulations allow an officer with a watchkeeping certificate to take the Mate's position on a vessel up to 3000 GT, and an officer with a Mate's certificate to take the Master's position on a vessel up to 3000 GT.

Ship's Masters must ensure that during the interim period of change, officers signing ship's articles are adequately qualified and in possession of valid certificates.

To this end officers who hold STCW '78, certificates will be required to revalidate their qualification to STCW '95 standards (see MSN 1692) by 31 January 2002.

Revalidation

Applications for revalidation of certificates should be made to the Registry of Shipping and Seaman at Cardiff (British certificates).

Applicants must have completed the standard short courses as required by the authority for the issue of a Class 4 or Class 2 Certificate of Competency and additionally show evidence of training and competence in specialized areas to show the holder can satisfy the additional knowledge, required by the relevant regulations.

Such training areas are identified as:

- GMDSS – General Operators Certificate
- Certificate of Proficiency in Survival Craft and Rescue Boats (replaces CPSC).
- Medical first aid and medical care.
- Radar (simulator) training.

Persons who are engaged in the control of fire-fighting operations are expected also to have completed advanced fire-fighting to meet the provision of section A-VI/3 of the STCW code.

All seafarers will also be required to meet medical standards of fitness as stipulated in Regulation 1/9, at intervals not exceeding 5 years.

Special requirements for specific ships

- Roll on-roll off passenger vessels – Masters, officers and other designated personnel will be familiar with crowd control and emergency situation management.
- Tanker vessels – Officers and ratings responsible for cargo operations must complete approved fire-fighting and tanker familiarization courses. Specialized training is required for all persons engaged in cargo operations for chemicals or gas tankers appropriate to the vessel on which they serve.

The route to certification: STCW '95

Every seafarer for both deck and engine room certificates must complete approved training to meet the standards of competence set by the STCW convention. Seafarers currently have the option of two methods to gain accepted qualifications:

1. The VQ system which is replacing previous methods.

2. The traditional non-VQ system for seafarers who obtained qualifications before the introduction of the VQ system.

The VQ system (for II/1)
New entrants to the industry wishing to obtain an STCW '95 watchkeeping certificate must have a minimum sea time of 12 months, of which 6 months must be spent on navigational watchkeeping duties under the supervision of a certificated officer.

To obtain the watchkeeping certificate, candidates must pass Merchant Vessel Operations (MVO) level 3, which includes all the ancillary short courses including Efficient Deck Hand (EDH).

In addition the candidate must also pass the oral examination set by the MCA.

NB. Other programmes may be approved by the authority with agreed levels of minimum sea service between 12 and 36 months if they are VQ based. (Sea service for non-VQ routes remains at 20 months.)

Alternative MNTB approved training programmes are already operational and in place for new entrant deck cadetships.

The VQ system (management level II/2), Chief Mate. A minimum sea time requirement of 12 months must be obtained by an applicant while holding an unlimited watchkeeping certificate (II/1) or a 1985 Regulation Class 4 Certificate.

NB. From September 1998 candidates for the Chief Mate's II/2 certificate do not require a Class 3 certificate from the previous regulations.

A candidate for the II/2 certificate will need to complete (MVO) level 4, and the underpinning knowledge as specified by the BTEC/Edexcel Supplementary Units in (a) Coastal Navigation and Emergency Response, and (b) Engineering and Control. (These two units provide the underpinning knowledge for level 5 VQ units as required for the Master (unlimited).)

The candidate must also obtain the Chief Mate's oral examination set by the MCA.

VQ system to qualify as Master. An applicant for Master (unlimited) must have 36 months' sea time as a watchkeeper of which 12 months must be while holding a Regulation II/2 certificate or 12 months as Chief Mate (II/2).

In addition to the requirements for Chief Mate the candidate must be in possession of MVO level 5 units in:

- Control of overall safety of navigation.
- Control of overall response to emergencies.
- Engineering and control.
- Ensuring the vessel's compliance with operational legislative requirements.
- Ensuring the vessel's compliance with commercial obligations.

The candidate must also obtain a pass in the Masters (unlimited) oral examination, set by the MCA.

The non-VQ traditional system
In the event that a candidate is not following an approved planned training scheme he or she would require 36 months' sea time serving in a deck capacity (general – purpose ratings would require 48 months), of which 6 months must include watchkeeping duties.

In order to obtain the STCW '95 watchkeeping certificate the candidate must follow a college-based course which leads to Higher National Diploma (HND) Nautical Science Part 1. The candidate should also be in possession of all the ancillary short course certificates as stipulated by MGN 2.

The candidate must pass the Scottish Qualification Authority (SQA)/MCA written examinations in:

(a) Cargo Operations and Stability.
(b) Coastal Navigation.

The candidate will also be expected to pass the oral examination set by the MCA.

NB. The current SQA/MCA written examination and the documentation of sea time experience is currently under review.

The non-VQ route for the qualification of Chief Mate. Candidates require 18 months' sea time while holding an unlimited watchkeeping certificate, i.e. STCW Regulation II/1, or a Class 4 certificate.

The candidate would be required to follow a college-based programme which leads to HND Nautical Science Part 2.

The candidate would be required to take the BTEC/Edexcel supplementary units in the topics of:

(a) Coastal Navigation and Emergency Response.
(b) Engineering and Control.

He/she would be required to pass the SQA/MCA written examinations in the subjects of:

(a) Navigation
(b) Stability.

Table 10.1 STCW programme

Implementation date	Programme item
1 February 1997	Flag States must decide what level of crew certification will be required on their ships. Port States must determine the (STCW) documentation which will be required by ships visiting their ports.
1 August 1998	An IMO listing (known as the 'White List') of Maritime Administrations will be produced, whose training and standards are judged to satisfactorily meet the new STCW requirements.
	NB. Many countries are still some way off from implementing the new regulations and standards (1999).
1 February 1999	GMDSS is fully implemented worldwide.
1 February 2002	The STCW convention revision of 1995 will become fully implemented.

The candidate must also pass the oral examination for Chief Mate set by the MCA.

The Non-VQ system to qualify as Master. A candidate who applies for the Master (unlimited) certificate must have 18 months' sea time while in the possession of a Chief Mate's certificate STCW '95 (II/2 certificate) or a 1985 Regulation Class 2/1 certificate.

In addition the candidate must also pass the oral examination for Master (unlimited) as set by the MCA.

The above methods of obtaining seagoing qualifications are as stipulated by the MCA for the UK. Other national authorities will have their own equivalent systems in operation which will satisfy the standards of required competency described by the STCW convention.

Similar infrastructures to the Deck Officer procedures are being established for the examination and certification of Engineering Officers. A Marine Guidance Notice is expected to provide further details in the near future. Table 10.1 shows details of the STCW programme and an example of the route to achieving Deck Officer's certificates is shown in Figure 10.13.

Application for certificates of competency

In the case of all applications for all grades of certificate by the non-VQ method candidates must show their sea service recorded/endorsed in the seafarer's Discharge Book and be supported by testimonials (watchkeeping certificates) signed by the Master of each ship served.

Dated qualifications

In some cases candidates may find that they hold a Class '2' grade certificate which is either:

(a) Dated before 01/01/1988 – if such is the case the candidate will be required to complete a college-based education course of two terms covering the subjects of Navigation Instrumentation, Engineering and Control, Business and Law, and pass external examinations in Ship Stability and Navigation.

Candidates who possess an HND or a UK degree in Nautical Science will only be required to follow an educational course of the examination subjects of Ship Stability and Navigation.

In every case the candidate must also undertake or the MCA oral examination.

(b) Dated after 01/01/1988 – provided it is obtained in accordance with Regulation II/2 of the STCW '95, or by revalidation in accord with Regulation II/2.

The candidate must undertake the MCA oral examination.

Additional reference on eligibility may be found in MGN 2 (M) and MGN 8 (M).

Appendix *Ship Master's Self-examiner*

Masters: legal

Q1 What are the Master's statutory duties in the event of a collision with another vessel?

Ans. Where it is practical to do so without incurring danger to his own ship and crew, the Master must:

(a) Stand by to render all assistance to the other vessel, her crew and passengers to keep them from immediate danger and remain with her until there is no need for further assistance.

(b) Exchange information with the Master or officer in charge of the other vessel, including the ship's name and the port to which she belongs, together with the name of the port from which she departed and the port for which she is bound.

(c) Report the collision incident to the Marine Accident Investigation Branch.

(d) Cause an entry of the incident to be made into the Official Log Book.

Q2 After collision with another vessel, what information and evidence would a Master collect for his owner's legal representatives?

Ans. The Master would be expected to complete a full report on the incident, and include any actions which may have led up to the accident being caused. He would be expected to collect all relevant log books, including the Engine/Deck Log Books together with movement books, instrument printouts, graphs, etc., recovered from bridge and machinery space instrumentation, e.g. echo sounder record, or course recorder.

It would also be considered essential to take statements from relevant witnesses, such as any pilot involved, or tugmaster, listing details of all vessels and parties involved with respective timings and positions of such parties.

The exact position of the incident must be stated and the respective navigational charts employed must be retained as evidence with all signals and communication reports from other shipping around or VTS authority in operation at the time of the incident.

A weather report of the area at the time of incident should be included in the Master's report, together with a list of all navigational instruments engaged before and up to the time of impact. Notation on special circumstances, such as anchors or moorings being employed, would also need to be included.

NB. Any video or photographic evidence which can be produced before, during or after the incident is usually well received by inquiry teams. Also officers and crew should be briefed not to make statements to unauthorized persons without the express permission of the Master.

Q3 If a Master is to 'note protest', what is he doing and who is he addressing the protest to?

Ans. The Master makes a statement under oath to a notary public, magistrate or British consul, often referred to as a 'proper officer'. The statement would relate directly with an incident or occurrence during the voyage, which has resulted in loss or damage.

Protest is noted when the cause of the incident is beyond the control of the Master and must be supported by 'log book entries'.

Q4 What are the conditions supporting a salvage operation?

Ans. (a) The property must be in danger.

(b) The salvor must provide his services voluntarily.

(c) The service must be successful (based on the clause 'no cure, no pay' under Lloyd's Open Form '95) to substantiate a claim.

Q5   What would be the content of a Master's reply to an offer of salvage in the case where it is the intention to accept?

Ans.   'I (the Master's name) accept salvage services from (salvor's name) on the basis of Lloyd's Open Form (LOF '95) under the clause of "no cure, no pay", for the vessel (ship's name)'.

Q6   When can a distress signal be used?

Ans.   (a) If the vessel is in grave and imminent danger, or if another vessel or aircraft is in that condition and cannot send the signal for herself.
(b) If the ship (or another ship or aircraft) needs immediate assistance in addition to that already available.

Q7   Who has the authority to use an international distress signal?

Ans.   The Master or the officer in charge of the vessel.

Q8   When is the Master of a vessel relieved of his obligation to respond to a distress signal which he has received?

Ans.   (a) When one or more vessels have responded and have already been requisitioned, and further assistance is no longer required.
(b) If the vessel in distress informs the Master that his ship is no longer needed.
(c) If another vessel at the distress scene states that additional assistance is no longer required.
(d) Where it is considered unreasonable to attend. Possibly because the risk to one's own vessel would be grave or the distress is from a vessel out of viable range from the Master's own ship.

NB. Masters should note that if their vessels are deep drafted, and as such may not be able to approach a vessel in distress because of reduced underkeel clearance, this may not be sufficient reason alone to relieve the vessel of responding under her legal obligations. Ships will undoubtedly carry a rescue boat or lifeboats with shallow draft that could possibly intercede without exposing the mother ship to grave danger and as such may well be capable of rendering assistance.

Q9   On receipt of a distress message what are the duties of the Master?

Ans.   The distress must be acknowledged and recorded in the communications log book of the vessel. The action of responding, or the reason for not responding, should also be recorded.

Q10   It is illegal for the Master to take, or any other person to send, a vessel to sea if it is in an overloaded condition. What action should the Master take if he found that the ship had been accidentally overloaded?

Ans.   The overload condition needs to be rectified by discharge of ballast, fresh water, feed water, or fuel. Once the vessel is at a correct state of loading the stability should be recalculated to take account of the changes made.

NB. As a last resort cargo would also have to be discharged, but this action may generate additional problems from shippers, consignees, Customs and owners.

Ship Master's business

Q1   What information would you expect to find on the front cover of an Official Log Book.

Ans.   (a) Name of the ship, Port of Registry and the official number.
(b) The gross and net tonnages of the vessel.
(c) The name of the ship's Masters together with their respective certificate numbers.
(d) The name and address of the ship's owners or the registered managing owner.
(e) The date and place that the log book is opened and closed.
(f) The official stamp and date when it is surrendered to the Registrar General of Shipping and Seamen.

Q2   What line of action would the Master take in the event that three crew members complain about the food or water?

Ans.   The Master must investigate the complaint and take such steps deemed necessary to rectify the problem.
He must also cause an entry to be made into the Official Log Book, to include:

(a) The names of the seamen making the complaint.
(b) The nature of the complaint.
(c) A statement detailing the action taken to rectify the problem.

The entry must also include a record as to whether the seamen who made the complaint are satisfied with the action taken. Arrangements must also be made for one of the seamen to further the complaint to a Marine Superintendent for further investigation if he/she so wishes.

Q3 What are the advantages of having ship's registration?

Ans. The Certificate of Registry identifies the vessel and provides the following:

(a) Protection from the flag state.
(b) British ownership carries a reputation of competence.
(c) As the owner of the vessel a mortgage may be raised on the ship as collateral.
(d) Proof of the Master's authority.
(e) Represents evidence of title of ownership (although not conclusive).
(f) The vessel will acquire the reputation of being a 'known vessel' which could help towards ship's clearance and Customs operations.

Q4 Describe what policies are addressed by the International Safety Management code and who is responsible for the development and implementation of this code?

Ans. The ISM code must address 'safety at sea', 'accident prevention' and avoidance of 'damage to the environment'. It is the responsibility of the company to develop and implement a Safety Management System aboard company vessels.

Q5 In the event of finding a stowaway on board the vessel after departure from a port, what action would the Master be expected to take?

Ans. Incidents which involve a stowaway should be treated as humanely as possible and in compliance with the UN convention relating to the status of refugees.

The Master should make every effort to determine the identity and nationality of the stowaway and attempt to learn at which port he/she embarked. The incident should be documented with all the known facts and reported to the authorities of the known port of embarkation. The owners should also be informed together with the authorities of the flag state at the next port of call.

Unless repatriation has been arranged with relevant documentation, the Master should not depart from the planned voyage in order to disembark the stowaway. All measures should be taken with regard to the health and welfare of the stowaway until the person can be allowed to disembark and be handed over to the correct authorities.

Q6 How could the Registrar of Shipping ensure that the medical standards of the ship are adequate, prior to allowing the Master to open the ship's articles.

Ans. The Master would schedule the medical locker to be inspected by a pharmacist before visiting the shipping office. The medical locker would be restocked and the Master would receive a pharmacist's certificate to this effect, which the Registrar would sight as acceptable evidence.

Q7 Under what conditions can the shipowner exclude his liability in total in respect of claims against damage to cargo carried on board?

Ans. On the understanding that the shipowner did not generate the reason for the claim, he is not liable for:

(a) Any loss or damage of the property on board if caused by fire.
(b) The loss or damage to valuables on board if the nature and value of the goods were not declared at the onset and listed on the bill of lading.

Q8 Under the ISM code, what two documents must be produced on demand to an inspecting 'proper officer' regarding the safe management of the vessel?

Ans. A copy of the original Document of Compliance (DOC), and a Safety Management Certificate.

Q9 What is contained in the Register of Lifting Appliances and Cargo Handling Gear?

Ans. The Register is maintained under the requirements of the ISM code and contains all the test certificates of wires, shackles, chains, hooks, derricks, cranes, davits, etc.

Q10 If a seaman is left behind in a foreign port, what entries would be made by the Master in the Official Log Book?

Ans. The seaman's name, and the date and the place where he is left would be entered into the log book. If the reason for his being left is known, this would also be included, together with any provision made for him by the Master. The employer (company) would also be informed of the incident.

Statements in the Official Log Book would be entered by the Master and witnessed by a member of the crew.

Emergency procedures

Q1 When navigating 10 miles off a 'lee shore', your vessel suffers a loss of main engine power. What action would a prudent Master take?

Ans. The Master must take the 'con' of the vessel and obtain an immediate position of the vessel and issue the following orders:

(a) Display NUC lights or shapes.
(b) Instigate a damage assessment and generate repairs if possible.
(c) Obtain an updated, current weather report for the area.
(d) Open up communications following an interim report on the damage.
(e) Establish a bridge team to include lookouts, standby helmsman, radar observer and Watch Officer.
(f) Estimate the drift rate under current conditions.
(g) Walk back both anchors clear of 'hawse pipes'.
(h) Update communications to Coastguard via coast radio station – an urgency call or distress call would depend on circumstances.

NB. In the event that on-board repairs cannot re-establish main engine power, communication with tugs must be considered as soon as possible.

(i) Inform owners of situation and conditions.
(j) Ballast the fore end of the vessel to weather vane the ship and so reduce the windage and the rate of drift.
(k) Consider use of a jury sea anchor to slow the drift rate down.
(l) Inform the MAIB.
(m) Cause relevant entries to be made in log books.

NB. It is not possible to provide an exact procedure for every different type of ship and every changing circumstance. Clearly, when compared with a general cargo vessel, a tanker vessel finding itself in this situation may decide to instigate communications sooner rather than later.

Other ship types could also generate a need for additional precautions, e.g. chemical carriers may create a need to evacuate shoreside positions because of the danger from toxics or other harmful products. Such actions would require early communications to move operations along.
    Passenger vessels may see the need to evacuate passengers in calm waters, rather than run the risk of taking the shore with passengers aboard. In the case of such an incident happening to a passenger vessel the Master is expected to make an immediate distress call, rather than an urgency call.

Q2 Your vessel has run aground in poor visibility, what line of action should the Master take?
Ans. The Master should take the 'con' of the vessel and instigate the following actions:

(a) Stop engines and sound the general alarm.
(b) Order the Chief Officer to make an interim damage assessment to assess watertight integrity of the hull, engine room condition (wet or dry), casualty report and evidence of pollution.

NB. The Master should advise the Chief Officer to specifically sight the condition of the collision bulkhead and the tank tops, and to obtain full soundings external to the hull as well as full tank soundings inside the vessel.

(c) The Watch Officer should be ordered to obtain the ship's position of grounding and to obtain the state of tide, whether rising or falling. Heights of high/low water together with the times of the next tides should also be obtained.
(d) The vessel should display signals for a vessel aground and commence sounding the appropriate fog signal if poor visibility remains.
(e) A full chart assessment should be made to assess the extent of the shoal, nature of bottom, available depth, nearest Port of Refuge, etc.
(f) Once the findings of the interim damage assessment are known, together with the position, the Master should open communications with the local authority via coast radio station. Depending on circumstances an urgency signal may be transmitted, or if the vessel is in danger of slipping back into deeper water with known hull damage, then a distress signal could be more appropriate.

NB. Fatalities, death related injuries, or a passenger vessel would probably provide an immediate cause for a distress transmission.

Additional communications should inform owners, charterers, agents, P & I Club, etc., together with the MAIB and/or the MPCU. An update on the weather report should be obtained and if refloating is being considered then tugs and surveyors would need to be contacted.

General statement following a grounding incident

No text can be expected to provide an exact format to cover every incident of this nature but a general outline of appropriate actions must be considered as a guide towards resolving the immediate situation.

The need to carry out a more detailed damage assessment to include internal and external soundings will become a necessity once the initial inspection has taken place. (The first damage assessment being carried out to allow essential communications to be made, as well as to provide an immediate overview of the situation.)

Where pollution has occurred an alternative line of action with extensive communication to the Marine Pollution Control Unit (MPCU) would be anticipated. Activity in line with the 'SOPEP' would expect to preserve life and property, while at the same time prevent escalation of the incident.

Where refloating is envisaged this would depend on what damage control actions have been possible and if the tank tops and the collision bulkhead have remained intact. Inspection by surveyors may be required before attempting to refloat and in any event a standby vessel should be in attendance prior to moving the vessel's position.

If the vessel grounds on a rising tide and is in danger of refloating accidentally a necessary action may warrant pushing the vessel further onto the ground, especially so if tank tops are known to be ruptured. Alternatively, adding extra ballast to the vessel to hold the existing position could also be an option to prevent the vessel with hull damage falling back into deeper water.

Where tanks are ruptured, water pressure could force oil from double bottom tanks upwards through sounding and air pipes to eventually lay on the uppermost deck. To this end Masters should order the deck scuppers sealed to reduce the pollution effects.

The saying 'never run ashore with an anchor in the pipe' is appropriate if time allows the use of anchors before the grounding incident takes place. In the majority of grounding incidents anchors may need to be deployed by carry-out procedures at a convenient time to assist with refloating.

Q3 Following the decision to beach the vessel after a collision incident, an oil tanker is observed to have ruptured tanks and is causing pollution. What actions would be expected by the Master to prevent escalation of the oil pollution.

Ans. The Master would expect to work towards the recommendations of SOPEP, and in any case of pollution make immediate communication through the emergency contact numbers. (This would effectively involve the shipowner and

cause back-up and support activity to be set in motion.)

In conjunction with senior officers, the Master would instigate emergency repairs, if possible, in the damaged and affected areas of the vessel. It would be practical to order an internal transfer of oil to take place from damaged tanks into empty undamaged tanks and so prevent more oil being lost overside provided that such undamaged tanks are available to do this.

Communication requests for lighter barges to enable an external transfer of oil to take place and for barrier/boom gear to be brought in should be made if the ship does not carry its own oil barrier equipment.

In the absence of designated barrier equipment being made available the use of floating mooring ropes stretched around to encompass the damaged hull could be seen as an attempt at containment which may stand the Master in good stead at any future enquiry.

Any action that hastens the clean-up activity must be considered worthwhile and to this end a request for 'Skimmer' vessels and /or quantities of dispersal chemicals may prove to prudent.

NB. The use of some chemicals which may cause additional environmental damage may not be allowed to be used without specific permission from the state authority affected, and Masters would be well advised to have such permission established prior to use of such commodities.

During any pollution incident, what might seem minor or every day activities could make a major difference to the outcomes long term. Efficient ship-keeping practice, such as no smoking on deck, and the accommodation doors being kept shut, could prevent casualties. Securing and sealing the uppermost continuous deck could, depending on the circumstances, reduce the amount of oil being lost overside.

Masters are expected to report any pollution incident and the Marine Pollution Control Unit (MPCU) provide advice and recommendations to help the situation. The Master would also cause entries to be made in the Official Log Book and the oil Record Book.

Q4 When at sea on a general cargo vessel, smoke is sighted coming from number 2 hatch, what activities would you anticipate to be ongoing aboard the vessel?

Ans. Once the smoke is sighted, the fire alarm should have been sounded by the Officer of the Watch and the Master would immediately take the 'con'.

The Master should issue the following orders:

(a) to the OOW – Obtain and chart the ship's position, update the weather report and open up communications with the coast radio station.
(b) to the Chief Officer – Take charge of the fire party and tackle the fire immediately. Keep the Master informed of all activity, close off ventilation to the space affected and commence boundary cooling of surrounding areas close to the fire.
(c) to the Navigator –

   (i) Plot a course to the nearest 'Port of Refuge'
   (ii) Plot an alternative course with the current wind direction from astern.
   (iii) Obtain and plot the position of the nearest ship to own ship's position.

(d) to the Chief Engineer – Depending on the circumstances affecting the fire the Chief Engineer should be ordered to place his engines on standby and the vessel's speed reduced to manoeuvring. He should further be ordered to ensure that the CO_2 total flood system was made ready to inject into number 2 hatch.

NB. The total flood CO_2 system is set in a state of readiness to inject into the engine room. In the event of a hold fire, the valve to the distribution line of CO_2 into number 2 hatch would need to be opened.

The Master should confirm his position to shoreside authorities and declare his intentions. It should be anticipated that an urgency signal would be despatched, or even a mayday, depending on the severity of the fire.

The Master should in any event order the NUC signals to be displayed and cause relevant entries into the Deck and Official Log Books.

A bridge team should be placed *in situ*, and the vessel placed on manual steering, communications and fire conditions being continually monitored.

Q5   While working cargo alongside in Hong Kong, the Master receives a radio message that a tropical revolving storm is moving towards the harbour position. What options are open to the Master? If he decides to remain tied up to the quayside, what security measures would he expect to take?

Ans.   The Master has limited options to ensure the safety of the ship:

(a) Let go and proceed outward to open sea (decision must be made early).
(b) Move the vessel to a storm anchorage (Hong Kong does have such a storm anchorage).
(c) Remain in port (not the ideal option unless a sheltered haven).

In all the above options, the ship would be expected to stop working cargo and eliminate any free surface effects to improve the stability condition. The Master should also order that the position of the storm is monitored by a continuous plot no matter what option is taken.

If circumstances dictate that the vessel has to remain in port the Master should cease cargo operations immediately and prepare the vessel for encountering severe weather.

Extreme weather calls for extreme actions and these could include all or some of the following:

(a) Close up and batten down all hatches. Seal the upper most continuous deck but ensure that scuppers and freeing ports are clear to shed any slack water.
(b) Request a clear berth away from dockside cranes and other similar quayside equipment. (Alternatively request dockside cranes are moved clear of the ship's vicinity.)
(c) Place main engines on stand-by and have full engine power available before the storm approaches.
(d) Request tug assistance to carry out and lay both anchors.
(e) Double up on all moorings to secure the ship with as many lines as possible. A variety of bights, long drifts and double springs might be considered appropriate.
(f) Have additional fenders rigged between the hull and the quayside.
(g) Ensure that all derricks/cranes are secured and that all cargo lashings are resecured, especially on deck cargo items.

Q6   When navigating in high latitudes the vessel starts to experience ice accretion. What action is the Master expected to take, if the ice growth threatened the ship's stability condition?

Ans.   Depending on the geographic position of the vessel the Master should consider the possibility of rerouteing to warmer latitudes. In any event the ship's speed should be reduced immediately to reduce the wind chill factor affecting the superstructure of the vessel.

Where the vessel is experiencing subfreezing air temperatures the Master is expected to make a statutory report concerning the prevailing conditions.

Ice concentrations should not be allowed to increase and the Master should order the Chief Officer to engage the crew to clear ice from upper deck areas. This could be achieved by steam hoses or breaking ice with the use of axes. Masters are advised that this activity is extremely hazardous, especially if ice concentrations have accumulated in upper rigging. Crew members should be briefed accordingly and wear adequate protective clothing when engaged in such activity.

Adverse stability, due to ice accretion, can also be bettered by improving the ship's ballast condition to counter any additional weight accrued from ice weight on the ship's upper works.

Ice accretion can also affect the ship's navigation instruments and the overall performance of the vessel while on passage. Aerials can often be broken due to heavy ice formations causing overload. Radar scanners may ice up causing reduced radar energy both outward and inward, effectively reducing the target definition being displayed.

Q7 While in command of a general cargo vessel at sea, you are called to the ship's bridge following the sighting of a distress signal from a small fishing vessel on fire and sinking. On approaching the craft a crew of five persons is seen in the proximity of the forecastle. How would you propose to recover the distressed men if the weather was Beaufort force '7', increasing to gale '8', and the position of the distress is outside the normal range of helicopter assistance?

Ans. The Master should take the 'con' of the vessel and obtain a detailed report of the event from the OOW, from the time of original sighting to the current situation. He would order a statement to be made into the log book and maintain a movement/incident record of all activity.

The prudent Master should establish a full bridge team to take station and place the engine room on standby, proceeding at best possible speed towards an upwind close-up position to the distressed vessel.

Consideration of several options should be examined:

(a) Launch own rescue boat (but with existing weather over force '6', this would probably not be viable without considerable risk to own ship's boat crew).
(b) Rocket line/messenger/hawser contact with the fishing craft should be established and

use made of their liferaft for transport, pulling it towards own ship. Survivors should board by gangway or scrambling net inside the lee of own vessel.
(c) Establish contact as in (b) and use own ship's liferaft on an endless whip to pull across to the distress ship and tow back with casualties on board.
(d) Manoeuvre own vessel alongside away from the burning section of the distress vessel. Ideally making contact with the fishing vessel forward of own collision bulkhead so reducing the likelihood of damage to own vessel.

NB. The freeboard of own vessel would normally expect to be higher than that of a small fishing craft and survivors could make use of 'Jacob's ladders' or nets over the bow of the high freeboard vessel.

Q8 Your vessel is in a mid-ocean position when the Chief Engineer reports that the stern tube is leaking. What recommendations and actions is the Master expected to order?

Ans. The Master must be primarily concerned with the possible increased level of flooding which may take place if the problem is left unattended. To this end he would order that the watertight door to the stern tube area is left in the closed condition (other than for access for maintenance purposes.)

It would be considered a prudent act to reduce the trim of the vessel and improve the overall stability of the vessel if possible, e.g. eliminate free surface and increase GM.

Order the repacking of the stern gland stuffing box if possible and establish a primary and secondary pumping arrangement in the affected compartment.

Investigate safe port options and inform owners of the need to effect repairs. Engine movements should be reduced if possible especially in the astern direction.

In the event that pumping arrangements are not holding the situation, additional strengthening for adjacent bulkheads must be considered. Communications may necessitate an urgency signal.

Continue to monitor the ingress of water levels by frequent soundings.

Q9 When bound for a port with a narrow channel approach and your vessel is known to have unreliable steering gear, what precautions would the Master take prior to entry?

Ans. No Master is expected to expose his ship to a potential hazard and as such it would be prudent to inform the company of the situation and take the vessel to a suitable anchorage close by, instigate repairs and fully test the steering gear prior to entering the channel approach. In the event that the company orders the vessel to enter the Master should take several basic precautions:

(a) Brief all bridge and engine room personnel on the situation and carry out a steering gear failure drill.

(b) Inform the port/harbour control of the suspected problem.

(c) Have an anchor party on continuous standby with both bow anchors ready for immediate use, once the vessel is underway and making way, towards the approach channel.

(d) Display 'Not Under Command' lights or signals.

(e) Inform the pilot on boarding of the steering problem and the status of the vessel.

(f) Request for additional tug support for braking and bow control.

(g) Station extra engineering personnel within the steering compartment.

(h) Reduce the speed of the vessel below what would be the norm for such an approach.

(i) Enter statements into the log book of all additional precautions taken.

(j) Additional intership communications should also be considered and the use of flag 'D' in the international code should be displayed.

Finally brief the helmsman to be on the watch for lack of response or any abnormal overcarry of the helm. Any additional problems being encountered or experienced should be notified to the bridge team immediately.

Q10 When on route to participate as a rescue unit in an SAR operation, what deck preparations would a Master order the Chief Officer to carry out.

Ans. Deck preparations depend on the target definition of the distress situation, and the number of expected casualties. However, the following actions would probably be required in most incidents:

(a) Turn out the rescue boat and have the rescue boat crew ready to launch.

(b) Prepare the ship's hospital (or suitable accommodation) to receive casualties.

(c) Place the Medical Officer and first aid party on standby ready to treat casualties.

(d) Turn out the accommodation ladder, weather permitting.

(e) Rig a guest warp overside in a compatible position to the accommodation ladder.

(f) Display search signals FS1 of the international code if appropriate.

(g) Rig scrambling nets overside if appropriate.

(h) Establish lookouts in high, prominent positions.

(i) Check ship's pyrotechnics, e.g. line throwing and 12 bridge rockets.

(j) Check and establish the readiness of the Aldis signal lamp.

(k) Test search lights and overside lights in readiness for night recovery operations.

Bridge operations

Establish the bridge team *in situ* and proceed at best possible speed to the position of commencing search pattern (CSP). Manual steering should be engaged and the Master should take the 'con'. The engine room should be placed on standby and an updated weather forecast obtained.

Close and continuous position fixing should be ongoing together with a continuous radar watch being maintained. A designated communications officer should maintain an effective watch under GMDSS requirements to OSC/MRCC or other search units. All activity and relevant communications should be entered into the log book together with positions of deviation being noted. (DF, if carried, could be gainfully employed for obtaining bearings on transmitting targets.)

Shiphandling and manoeuvring

Q1 What are the dangers when a tug is drawing alongside the parent vessel in the bow region, prior to accepting a forward towline?

Ans. The main concern for the tugmaster and the Master of the mother vessel is interaction. As the tug moves alongside it is expected to generate a pressure cushion between the two hulls. Such pressure is greatest at the shoulder, near the break of the foc'stle, and causes an outward turning motion on the smaller tug vessel. The tugmaster needs to counter the outward turning motion by applying helm towards the parent vessel.

The danger is that the tug, now with helm applied, is still moving forward to obtain the correct station. As the tug moves beyond the shoulder of the parent vessel the pressure under the flare of the bow decreases dramatically and the tug still carrying helm towards the ship is drawn in towards the bow of the larger vessel, with the possibility of a collision.

Q2 When letting go a tug from the aft station of the vessel, what engine movements would the Master anticipate if the tug is secure by your own ship's rope through a centre lead?

Ans. The main concerns at the aft end of the vessel are for ropes/wires in the water and fouling the propellers. The situation can be relieved by having an efficient officer on station who is aware of the dangers of the combined use of mooring ropes and propellers.

In this case, because the tug is secured by a ship's rope and the rope will float, the Master would order half or full ahead on main engines, after giving the order to 'let go' the aft tug. The stern wake would trail the towline well astern and on the surface once released from the tug. This provides the officer on station ample time to take the rope to the winch and heave it aboard, without risk of fouling the propeller.

Q3 Referring to Figure App. 1 state how you would berth your own vessel starboard side to, between two ships already secured alongside. Your own vessel is in a position (1) stemming the ebb tidal stream. A mooring boat is available, and the vessel is fitted with a right-hand fixed blade propeller.

Ans. (a) The vessel must be manoeuvred to stem the tidal current as in position 1.

(b) Manoeuvre the ship to position 2 parallel to the moored vessel A.

(c) Run the best ship's mooring rope from the starboard quarter to the quayside with the aid of the mooring boat 2 and keep this quarter rope tight, above the water surface.

(d) Slow astern on main engines – then stop. This movement should bring the vessels stern in towards the quayside by means of transverse thrust. The bow should therefore move to starboard, outward to bring the current onto the ship's port bow – position 3.

(e) The vessel should turn with current effective on the port side and no slack given on the quarter rope to complete an ebb swing – position 4.

(f) Run the forward head line (position 5) and draw the vessel alongside from the fore and aft mooring positions.

NB. Where the manoeuvre is required when an offshore wind is present, use of the offshore anchor reduces the rate of approach towards the berth.

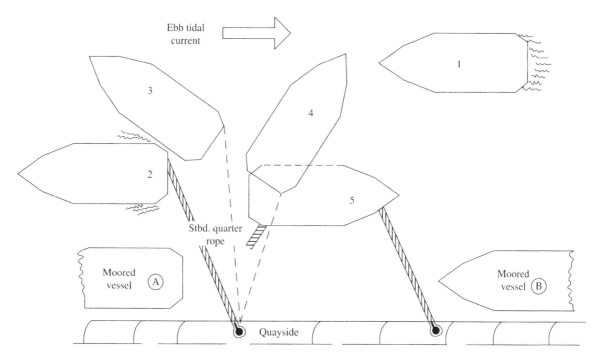

Figure A1

If an onshore wind is present the first bow line would probably need to be carried aft to allow it to be passed ashore or alternatively use the same mooring boat which was initially employed.

If no mooring boat was available for berthing, the quarter line could be passed, first to the moored vessel A, and then onto the quayside.

Q4 A vessel at anchor observes another vessel on a collision course heading towards her. What options are open to the Master and which of these options would it generally be more prudent to take?

Ans. Depending on the circumstances of the case, the vessel at anchor would normally be expected to maintain anchor watches and because the vessel is still considered at sea, should have kept her engines on 'Stop'.

The following three options available to the Master should all include trying to attract the attention of the approaching vessel by sounding five or more short and rapid blasts on the ship's whistle supplemented by an (Aldis) light signal directed towards the bridge of the oncoming vessel.

(a) Pay out more scope on the anchor cable and allow the vessel to drop astern out of the line of approach of the oncoming ship. (Probably not the best option as an approaching vessel, if it finally decides to alter course, would probably move to pass astern of the anchored vessel, clear of the cable region.)

(b) Heave up on the anchor cable to reduce the scope and pull the anchored vessel ahead out of the line of approach.

NB. This option involves taking the time to get an anchor party mustered and ready at stations to work anchors and cables and would need to reflect the range of an approaching target vessel.

(c) Steam over the anchor cable using main engines. This would have an immediate effect of taking the vessel out of the line of approach. It does not rely on members of an anchor party to respond and should effectively place the vessel quickly into a safer position.

This must be considered by far the best of the three options on the assumption that engines were left on 'Stop' and at immediate readiness.

Q5 When a vessel is 'hove to' at reduced speed, what are the advantages of prudent ballasting?

Ans. Prudent ballasting can improve the overall stability of the vessel and lend to a more seaworthy condition. Depending on the state of loading it could immerse the propeller deeper and provide better speed/manoeuvrability. Correct ballasting could also reduce stress levels and ease any 'pounding' that the vessel is experiencing. Draught and trim conditions can be improved to provide a more comfortable motion to the vessel.

Q6 When navigating in close proximity to high speed craft, e.g. hovercraft, hydrofoils, etc., what special manoeuvring characteristics would a Master be cautious of?

Ans. Fast moving vessels may not have the ability to display navigation lights in the more recognizable format because of their overall construction. Neither can the other vessel's sound manoeuvring signals be readily detected over and above the noise of their own engines.

Fast moving craft generally have great manoeuvrability and can be plotted on radar; however, plots change rapidly and radar observers should be aware of increased speed levels resulting from completed plots.

The majority of these vessels are equipped with powerful bow and stern thrusters. When docking sharp turning motions and tight berthing procedures could well be the order of the day, often supported by local navigation warnings from harbour control units. Other vessels should endeavour to give a wide berth when these craft are so engaged.

A distinct difference between a deep draughted tanker and a hovercraft is that one has restricted sea room because of draught. Whereas the other has virtually no draught restriction whatsoever. Masters should be aware that certain fast craft can use this asset when involved in collision avoidance and as such may take a course of action that might at first appear a dangerous or impossible manoeuvre.

Q7 When on passage through a traffic separation scheme and following the general direction of traffic, a trawler is seen off the port bow on a steady bearing with range closing, just leaving the central separation zone. Assuming it is daylight and visibility is good, what would be the suggested actions/options open to a Master of a general cargo, power driven vessel, if it is clearly established that the fishing vessel is going to cause an impediment to his vessel's passage?

Ans. The fishing vessel is directed by rule 10 of the COLREGS, to avoid impeding the passage of a vessel following the lane. As such she would appear to be contravening this particular regulation. However, it is the duty of the power driven vessel to keep out of the way of the trawler under rule 18 of the COLREGS.

The possible actions open to the Master are to sound five or more short and rapid blasts on the ship's whistle to express his doubt as to the continued intentions of the trawler. As the incident is occurring during the hours of daylight, the Master could also order the display of the international code flags YG. If no response is made by the fishing vessel to these sound and visual signals the Master must take such action as will best avert collision. This could be a reduction of speed in ample time to allow the trawler to pass well ahead of the vessel. The Master would be within his rights to take the name/port of registry of the fishing vessel and report the incident to the Coastguard.

Alternative action would be to carry out a chart assessment to ascertain the available depth of water outside the traffic lane. If adequate underkeel clearance is available, an alteration of course to starboard out of the lane should be made. Circumstances may make a complete 'round turn' necessary to re-enter the separation scheme. Such a manoeuvre would need to be carried out in ample time and should not bring the vessel into another close quarters situation.

Of the two options to avoid the impeding trawler, the reduction of speed in ample time would appear to be the better action. It keeps the power driven vessel inside the TSS limits, it should not compromise the vessel's underkeel clearance and complies with the regulations.

Q8 Your vessel is expected to proceed in an ice convoy, what manoeuvring details would the Master expect to clarify prior to rendezvous with the other ships in the convoy?

Ans. The Master could expect to establish communication with the Commander of the ice breaker leading the convoy and establish the rendezvous position.

The Master should pass on relevant details regarding his own ship's manoeuvring data and expect to be informed of the following information:

(a) The ship's position inside the convoy order.
(b) The expected operational speed of the convoy (this is dictated by the speed of the slowest vessel).

(c) The separation distance between the vessels in the convoy (this is sufficient to avoid collision and at the same time provides benefits from the wake of the lead vessels, probably between 150 and 200 metres).
(d) Communication channels and special signals operational for ships in the convoy.
(e) Contingency plans in the event of not being able to continue, i.e. recommended anchorage positions.
(f) What type of towing arrangement will be required in the event of becoming beset in ice and towing becomes necessary.
(g) Current ice report on the type, coverage and thickness that can expect to be encountered at various stages of the passage, if known.

Q9 A roll on-roll off vessel equipped with bow thrust and twin controllable pitch propellers must berth port side to use the stern ramp facilities.

How could the vessel be turned and berthed with the reduced swinging room available between the quayside and the shoals? (See Figure App. 2.)

Ans. (a) Vessel approaches at slow speed ahead from position 1.
(b) Starboard anchor let go as if to snub the vessel around 2.
(c) Cable allowed to run before being checked. Helm hard to Port.
(d) Bow thrust engaged 100 per cent, cable checked, port propeller ahead pitch, starboard propeller astern pitch and allow the vessel to attain position 3.
(e) Use a mooring boat to run a mooring line from the starboard quarter to the 'dolphin' and secure at position 3.
(f) Use bow thrust and port propeller with ahead pitch to bring the vessel parallel to the quayside 4.

Q10 Your vessel is ordered to rendezvous with a helicopter for a 'land on' operation while passing around the area off Cape Town, South Africa. What aspects of manoeuvring should the Master consider as necessary considerations, prior to engaging with the aircraft?

Ans. The Master should call a meeting of relevant personnel directly involved with the landing of the helicopter. In particular he would be concerned with briefing the bridge team and ensuring that the engine room staff are fully informed of the pending operation.

An updated weather forecast should be obtained, and a full chart assessment of the

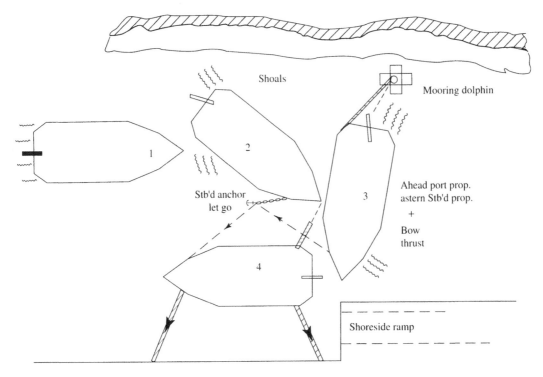

Shoals

Mooring dolphin

1

2

Stb'd anchor
let go

3

Ahead port prop.
astern Stb'd prop.

+

Bow
thrust

4

Shoreside ramp

Figure A2

rendezvous area made. Ideally such an area should have adequate sea room, with no restrictions affecting the vessel's underkeel clearance.

A designated track for engagement with the aircraft should be plotted, taking into account the wind direction. Such a track should be clear of traffic separation schemes, focal points, pilot roads and other similar congested areas.

Speed of the vessel on approach and during engagement should be established and passed with the vessel's heading to the pilot of the helicopter.

The Chief Officer would normally be delegated to prepare the deck area to engage the aircraft and this would include turning out the rescue boat.

NB. The Master should consider the weather conditions in relation to keeping the operation ongoing. In the event of an accident, necessitating the use of the rescue boat, the Master would be expected to account why he was unable to launch the rescue boat, and as such why the helicopter operation was not postponed to a time of better weather.

Manual steering should be engaged in ample time, prior to approaching the operational area,

the ship's engines placed at standby, and the vessel held at manoeuvring full speed. If the vessel is equipped with a bow thrust unit this should be kept readily available.

Correct signals showing 'Restricted in Ability to Manoeuvre' should be displayed once the vessel engages with the aircraft.

Miscellaneous

Q1 What are the special features usually present with roll on-roll off vessels?

Ans. The quick turnround of this type of trade necessitates that the ships are highly manoeuvrable with combinations of twin propellers (usually CPPs), twin rudders, bow and/or stern thrust units, stabilizers and occasionally bow rudders.

The operational features include strengthened decks with internal ramps to facilitate movement of vehicles to upper/lower deck levels. Such decks are fitted with extensive deck securing points to lash vehicles in position for the duration of the sea passage.

The vessels are extremely vulnerable to flooding and as such are usually built with good drainage facilities. There are no transverse bulkheads on vehicle decks, as these would prevent traffic movement, and the use of cantilever frames is widely employed to provide the required continuity of strength.

Vehicles gain access by either stern door/ramp or ramp/bow visor (see Figures App. 3 and App. 4).

Other special features could include any or all of the following: raisable car decks, internal vehicle lifts, anti-heeling tanks, trimming tanks, sprinkler fire-fighting systems, water curtains, CCTV to the bridge, watertight integrity tell-tales monitoring hull door/access points inclusive of shell doors.

NB. Crew members require specific training in crowd management and safety, crisis management, and hull integrity training for ro-ro passenger ships.

Q2 A vessel may be in need of a tow without necessarily being in distress, and the Master of another ship may find the circumstances such as to volunteer to take another vessel in tow. If such is the case what must the Master check and ensure prior to entering into an agreement of towage?

Ans. The Master should first ascertain that the charter party and the bills of lading do not prevent the vessel engaging in such a towing operation.

He would also ensure that the ship is capable of conducting the towing operation in a seaman-like manner. Is the ship equipped with adequate towing springs, for instance? Is the fuel capacity adequate to reach the port of destination, and will the speed of towing affect the condition of your own ship's cargo? (Perishables may prevent such an operation.)

The practical question of the ship's power being sufficient to manage the task would also have to be raised.

Assuming that the vessel has the capability to conduct the operation, the Master should expect the owners to confirm agreement and if under charter ensure that the charterers and underwriters are informed. (There would probably be an increased premium to pay.)

Obviously circumstances would influence whether you can engage in towing. Either way

Figure A3 Bow visor shown in the raised position on a roll on-roll off passenger vehicle ferry

Figure A4 Vehicle stern ramp deployed on a conventional roll on-roll off ferry, laying flush on the shoreside ramp. The vessel cannot be allowed to list or the ship's ramp will twist and not remain accessible to vehicles. Some of these vessels are fitted with automatic trimming/stabilizing tanks to prevent a list occurring, no matter what the ongoing loading condition is

the consideration should be made whether the value of the towed vessel, plus the value of its cargo, is worth the effort necessary to achieve a successful result. Bearing in mind that in order to make a salvage claim the operation must be deemed successful.

NB. The above statements refer to a vessel not in distress, and Masters should be advised that the obligations to assist a distressed vessel override any contractual agreements.

Q3   Distinguish the difference between open loop and closed loop control systems, as applied to a ship's steering gear.

Ans.   **Open loop** – If the helmsman is trying to steer a straight course, he will compare the measured value (MV) of the ship's head from the com-

pass, which in this case would be the sensor, with the desired value (DV) from the course board.

If these two values differ, then an error exists and the helmsman should apply a corrective amount of 'wheel'. The system itself does not ascertain the error, nor does it employ the error to initiate correction, hence the loop is an open one and requires the action of the helmsman to complete the cycle.

Once an error signal is generated by a difference between the desired value (DV) and the measured value (MV) automatic correction is made with the closed loop system. Instrumentation within the automatic steering unit continually monitors the difference quantity and applies a suitable corrective action to maintain the closed loop operation.

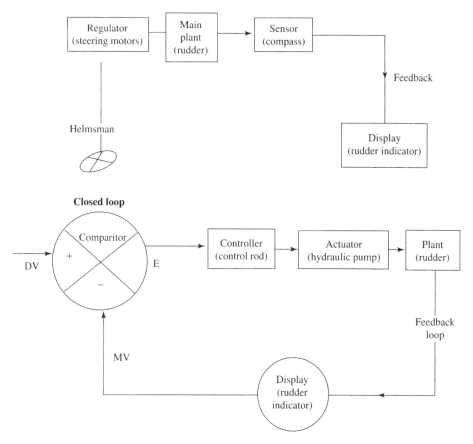

Figure A5 Open loop

Q4 What are the general requirements that a Master should pay particular attention to when loading a full deck cargo?

Ans. Deck cargo should be stowed and distributed with regard to the load density plan so as to avoid excessive stress which might affect the strength of decks and supporting structures.

The Master should also pay particular attention to the following points:

(a) Stability – that the vessel will retain adequate stability throughout the period of the voyage. That the loss of weight of water, fuel and stores through consumption will not adversely affect the stability condition.

That the possible increase in weight by the ship or the deck cargo from ice accretion or water absorption will not generate a dangerous condition.

That due regard is taken of any increased windage effect created by the vertical positioning of deck cargoes with regard to the expected prevailing winds for the voyage.

(b) Watertight integrity – the Master should ensure that the watertight integrity of the vessel is not impaired by the stow of deck cargo. Ventilators and air pipes should be clear and protected and all access points can be opened/closed as required.

(c) Operations – the deck cargo stow should be seen as not to interfere with the navigation or general operations of the vessel. Special note should be made of access to engine rooms and steering compartments and the visibility of watchkeepers from the bridge.

(d) Access – unless crew access is provided by a passage on or under the deck carrying the deck cargo a walkway must be constructed not less than 1 metre wide over the deck cargo. Such a walkway must be safe and efficient and fitted with guard rails or wires to a height of not less than 1 metre (minimum three courses of rails or wires).

(e) Securing – it is normal procedure to expect the Chief Officer to ensure that the cargo is well secured against all weather conditions.

Lashings should be plentiful in number and of adequate strength and the prudent Master should work with the Chief Officer to ensure a secure stow.

Q5 A vessel is to make an intended passage via the Straits of Singapore, a well-known region for attacks by pirates. What preparations and preventive measures could the Master take to avoid an attack on his ship?

Ans. Masters should realize that there are no effective methods of preventing the ship being hijacked against a well-organized, well-equipped and determined adversary.

Pirates normally go for light, easy, transportable goods and especially cash or jewellery. Approach and boarding of the ship is very often achieved by fast launch and boarders are more often than not armed.

Masters should combat attacks by briefing their crews to be continually vigilant and maintaining effective lookouts, especially in high risk areas. Masters themselves should be wary of false distress calls and try to navigate at maximum distance offshore (recommended 50 nautical miles).

The Master should recommend that the company fits anti-hijacking devices to the ship before departure. Current devices include a small transmitter 'SHIPLOC'. This is a satellite tracking instrument linked via the internet to a PC at a monitored station. Such equipment can generate an alarm to alert the authorities, especially maritime patrol vehicles.

Additional anti-piracy equipment could also include security cameras and long ranging overside lighting. Such measures will not stop the determined attack, but may generate time to signal an alarm and bring in patrol craft that much sooner.

An anti-attack plan should be devised to include evasive manoeuvres at maximum speed. Deck patrols should be maintained in high risk areas and in harbours. When berthed alongside, tight gangway security should be established to prevent unauthorized persons reconnoitring the vessel as a potential target. The crew should be well briefed on the alarm procedures and should be advised on any company policy if and when confronted with an act of piracy.

Q6 What type of response would the Master expect passengers and crew to adopt following a collision at sea and what type of phases would the casualties expect to pass through in order to be 'a survivor'?

Ans. Individuals being thrown into an emergency situation with or without death related injuries will cause a variety of reactions. Fear, excitement, shock, stress could all be involved creating both rational and irrational behaviour amongst those affected. Casualties could expect to pass through the following phases:

(a) The warning phase – A warning may or may not be given, and it would be incorrect to assume that all persons on board hear it, or if they do, that they treat it with serious intent. There will be doubt and a level of ignorance about what to do, leading to greater or lesser degrees of confusion. It should always be remembered that this phase may not happen at all, but if it does, the level of training and discipline amongst the crew could make a difference regarding overall survivability.

(b) Impact phase – Persons on board will hear it, see it, or feel it. Either way the psychological realization that something is wrong, that an emergency may exist or actually exists becomes fact. Behaviour patterns are unusual, and people react differently in extreme circumstances and in different settings (one doesn't expect to join a cruise ship and leave it in a lifeboat). This can be a brief phase, but one that can have the mixed emotions of fear and helplessness, coupled with a sense of realism and excitement. Personnel could be expected to do little or even carry out incorrect actions immediately following impact, while some will behave in a completely rational manner. At the same time others may become immobilized and not do anything at all.

(c) Abandonment/evacuation phase – With the need to escape becoming more and more apparent (survival craft being launched, persons leaving the ship, dense smoke visible, etc.), personnel will generally act often in a familiar manner, e.g. using an exit by which they entered, as opposed to the nearest way out.

The feeling of entrapment comes to the fore and desperation can easily surface. Passengers have a need to feel that they are being cared for and the functional officer, taking charge, acting in control can be a positive influence to move many people from a hazardous area to a relative position of safety. Passengers want to be told what to do, in the belief that whoever is doing the telling has the authority and knows what they are doing. The time factor is also critical during this phase, and an early, accurate assessment of any situation can only speed up the positive movement of personnel away from the danger area.

(d) The survival phase – The general circumstances and the very nature of the incident will have a direct influence on the period of time that the survival phase will last. Persons clear of danger, in a survival craft in good weather, will probably feel exhilarated to be away from the dangers of the parent ship. A similar survival craft in bad weather could experience a feeling of 'out of the pan into the fire', so to speak.

Anxieties can expect to be high as to whether a distress message was sent before or during the abandonment phase, and more to the point was it acknowledged? What weather conditions are present and how far from the nearest rescue authority is the parent vessel could all influence rational behaviour in the close confines of a ship's lifeboat.

Potential survivors need positive leadership within this survival phase, and any leader of the group will need exceptional management skills to keep control.

(e) Rescue phase – No one is a survivor until they have been rescued, as the saying goes. Any potential rescue vehicle will generate an overall feeling of thankfulness, but in reality, this is far from the case.

This is the one phase that discipline and common sense must be held together by the person in charge of the group. The behaviour might be reactionary when complying with necessary orders, especially so close to safety. Others might start to experience post traumatic shock or act out of character. A steady, firm and controlled manner becomes essential to complete this phase.

(f) Post trauma phase – Persons having successfully passed through the previous phases could start to experience symptoms of withdrawal while others just want to talk about their experiences to the point of talking to anyone who will listen.

Long term they may experience 'flashbacks' which could necessitate psychiatric help, or treatment for depression. This phase could last for an indefinite period of time, with or without help, depending on the level of experiences that the individual has witnessed.

Q7 What are the benefits and operational characteristics to the Master if his vessel is fitted with a bow thruster unit?

Ans. (a) The bow thruster unit has the capability to turn the vessel in its own length without having to resort to main engine power and use of rudder(s). Turning the vessel can also be achieved more quickly.

(b) The capacity of thrusters can usually hold the vessel on station, even in adverse and severe weather conditions.

(c) Thrusters, depending on type, can often be employed as an auxiliary source of power, especially so in the event of main engines being disabled.

(d) Effective use of thrusters can reduce or eliminate the need for tugs when berthing or undocking.

(e) Additional power needs to be generated to operate thruster units.

(f) To achieve effective use, remote positions of thruster control need to be established inside the bridge and on either bridge wing.

(g) Before starting thrusters, the surface area in the vicinity of the thruster should be checked for debris or ropes, etc. which could foul thruster propellers.

(h) Additional maintenance is required to keep thrusters operational. Maintenance, other than routine checks, usually becomes an additional dry dock task.

(i) A thruster may be useful when at anchor to ensure that the vessel swings in the desired direction.

Q8 What would you understand by the phrase 'load on top' when referring to an oil tanker vessel?

Ans. 'Load on top' is a system which was initially developed to save money, but it also became a system which benefited the environment.

When an oil tanker discharges her cargo, some residual is always left inside the tanks. These spaces must be cleaned of deposits before loading the next cargo. This was previously achieved by use of high pressure washing machines. The mixture of the remaining oil and water was then discharged overside.

Fortunately, this practice has generally now ceased, and the mixture is now allowed to settle, leaving an interface with the oil residues floating on top of the water (pumped into a 'slop' tank). The water can then be pumped away, via a separator, leaving only the oil remaining. Once at the loading terminal a fresh oil cargo can be loaded on top.

The financial benefit to the shipowner is that considerable amounts of what would otherwise be wasted cargo can be reclaimed, and the benefit to the environment is cleaner seas.

Q9   What are the services and objectives provided by a VTS system?

Ans.   A vessel traffic service (VTS) provides the following:

(a) Information/communication service.
(b) Navigational assistance service.
(c) Traffic organization service.
(d) Co-operation with allied services.

The objectives of a VTS operation are to improve the safety and efficiency of traffic and provide greater protection of the environment.

Q10   Describe the objectives and structure of the International Maritime Organization (IMO).

Ans.   The IMO is a specialized agency of the United Nations Organization which has a main objective of achieving co-operation amongst national governments on maritime safety and shipping, including marine pollution control and protecting the environment.

The objectives are attained by debate and agreement with the following:

- The Assembly (governing body) of representatives of all member states. The Assembly meets every 2 years to determine policy and plan a working programme. It also votes a budget and elects members to council and committees.
- The Council acts as IMO's governing body between sessions of the Assembly. It consists of 32 member governments.
- Committees – Safety Committee (most senior), Legal Committee, Technical Co-operation Committee and, Facilitation Committee.

Full supporting staff of international civil servants are established in London to service the organization.

Index

Printed and bound by CPI Group (UK) Ltd, Croydon, CR0 4YY

17/10/2024

01775697-0004